CONTRIBUTIONS OF SCIENCE TO RELIGION

BY

SHAILER MATHEWS

DEAN OF THE DIVINITY SCHOOL, UNIVERSITY OF CHICAGO

WITH THE COOPERATION OF

WILLIAM E. RITTER   ELLSWORTH FARIS
ROBERT A. MILLIKAN   CHARLES H. JUDD
EDWIN B. FROST   JOHN M. DODSON
EDWARD B. MATHEWS   CHARLES B. DAVENPORT
C. JUDSON HERRICK   E. DAVENPORT
JOHN M. COULTER   C-E. A. WINSLOW
HORATIO HACKETT NEWMAN

D. APPLETON AND COMPANY
NEW YORK :: LONDON :: MCMXXIV

PRINTED IN THE UNITED STATES OF AMERICA

5004

CONTENTS

PART II

SCIENTIFIC COOPERATION WITH NATURE

PART III

RELIGION, THE PERSONAL ADJUSTMENT TO ENVIRONMENT

CONTENTS

LIST OF ILLUSTRATIONS

CONTRIBUTIONS OF SCIENCE TO RELIGION

CHAPTER I

INTRODUCTION

A MAN'S religion must not give the lie to the world in which he lives. Yet, from the days of primitive savagery, religion and science have often been opposed to each other. Sometimes this opposition has been unqualified; sometimes it has aimed to divorce investigation by science from the faith of religion. It is rarely we find a person who has not in some way organized his life in at least two planes; in one of which he admits the operation of the law of cause and effect, and in the other relies on a supernaturalism varying from a confidence in rabbits' feet to a recognition of design in nature. Other contradictions have come in; on the one hand the scientific man is tempted to conceive of religion as still embodying the prejudices and hostilities of some early period, and on the other hand the religious man is sometimes vehemently opposed to any scientific formula which, like that of organic evolution, is not in accord with scientific beliefs he has incorporated into his theology.

Such irrational attitudes are explicable when we recall that in science and religion alike we are dealing, not with abstract truths, but with folks—with scientists and religious people. How far the methods and accomplishments of the one can help the other will be determined by a man's capacity to reconcile intellectual patterns of the one with those of the other.

Nor should it be thought that the difficulty lies in the human weakness of theologians alone. The scientist is also subject to prejudice and regard for authority. Indeed, not a little of the controversy between science and theology usually charged to *odium theologicum* has been due to the pertinacity, if not the pugnacity, of scientists in revolt against scientific views taught theologians by their predecessors. Much of "the conflict between science and theology" might with equal accuracy be described as the conflict between champions of scientific views admitted into religious thought and champions of other views not yet admitted.

What is religion? This is a question given many answers. Of late, however, the dispassionate study of human institutions has led to the discovery that within all peoples are social actions and attitudes having the same purport and functions. These actions might be described as efforts on the part of men to gain help and the goods of life by the aid of what they regard as superhuman personal forces in the environment upon which they are dependent. Of course, the understanding of environment varies with the culture of peoples. The bushman of Australia and the negro pigmy of central Africa could hardly be expected to have the same world-view as a university professor. Yet even they are representative of civilizations which are not absolutely primitive, for thousands of years separate them from the men of the Old Stone Age, to say nothing of still earlier human life.

The means by which these advantages from superhuman elements of an environment were to be gained have been various. Indeed, one cannot understand any religion of to-day without knowing the course of humanity's history.

Sometimes men have approached natural forces directly, treating them as animate objects; sometimes they have discovered spirits with good or evil intent in objects which they have made into fetiches; sometimes they have found in nature a supreme mind and purpose. They have organized rites, maintained customs, established feasts, offered sacrifices, made gifts and prayed to idols and unseen gods. Within each one of these general classes of activity there is something incalculable, yet the function of each is just as definite as that of the social activity which has developed political institutions.

A religion may be described as the complex of those social acts by means of which a group undertakes to ward off the anger and gain the help of those superhuman and not understood elements of its surroundings upon which it depends. Speaking generally, such acts reproduce those already practiced by the group in its social life and sanctified by age and long repetition.

But why should a whole group have built up institutions of a religious sort? An answer to such a question cannot stop with sociology. It carries one back into the depths of human nature. Religious practices of the whole group are the product of a desire of the members of that group for self-protection and advantage. The sense of need of help from more than human sources springs from a sense of man's own helplessness and dependence upon such sources. The river, the storm, the sun, even stones and pieces of wood seemed always liable to hurt or able to help. It is hardly a complete statement to say that fear made the gods, but it can hardly be doubted that need and mystery made religion. Men have treated these powers as they have treated contemporaneous persons of authority. At the start they had no philosophy or theology; they had only their own life and experiences and that power of imagination which is so universally the forerunner of knowledge. In the social practices constituting a religion we come upon a universal urge and attitude which are not mere reason but none other than a phase of life itself. A religion viewed thus inductively is seen to be a phase of the life process of some group in which life seeks to protect itself by an attitude toward the mysterious element in the environment, which embodies: (1) a sense of dependence upon the same, (2) attempts to gain therefrom help through the establishment of personal relations, (3) the utilization of cultural organization in the accomplishment of such attempt. So far a religion is the perpetuation of inherited group customs. It is a social inheritance which must be rejected or accepted.

To think of religion as a phase of the life process is not, however, to deny it intellectual elements. Certainly reason belongs to life as truly as impulse. As to the logical priority of each element in an attitude or an act, it is not necessary to decide. They are each and all present. So in religion. Unless men are convinced that personal relations with their environment are reasonable, they will soon abandon attempts to establish them. They are constantly questioning the reasonableness of their inherited religions. In consequence, whether expressed or not, religion implies some general working hypothesis as to man's place in the total scheme of things and, also, as to the scheme of things itself—that is, a more or less developed world-view. It is, in fact, one of two world-views logically tenable; the materialistic or naturalistic, and the religious.

1. The former was not infrequently held by the scientists of the nineteenth century and is still championed—sometimes rather dogmatically—by some of those of to-day. In its older form it predicated more or less explicitly a dead matter of which all thought and emotion were outcomes. Mind and matter were regarded as two fundamentally

opposite things, the first alone real. Mind was regarded as a superfluous supposition. Physical and chemical forces alone were discoverable.

The more recent type of materialism might better be called agnostic monism. It regards matter as energy of which mind is an expression. With such monism as a metaphysical postulate (or as its adherents would probably say, conclusion) religion might, and in some cases has, made terms. But the modern materialist goes further. He asserts that the uniformity of nature is hostile to any freedom of personality in the real sense; that there is no evidence of purpose in the universe (although he speculates as to "its goal"), that there is no form of existence other than those "envisaged by physics and chemistry."

Yet no materialist denies the existence of religion as a social fact. He simply denies its place in a rational scheme of things. And it is this denial that separates him from the religious man. To both alike the elements of scientific knowledge are available. Both alike must accept the universe as it is. But the materialist minimizes the human activities we call personal and reduces all knowledge to sensation. This is not science; it is philosophy or even much-berated metaphysics. Perhaps more truly can it be described as the application of a deductive logic to a major premise that is an untenable and sterile assumption. At all events, it must itself submit to the same tests that would apply to the religious world-view. Both alike must stand They must furnish a probable hypothesis for all germane facts.

For the great issues of every soul are not solved merely by logic or by knowledge. Suffering, disappointment, fear, bereavement, old age, death, are not on the plane of science. The death of a Pierre Curie, run down by a truck, is not explained or relieved by a knowledge of radio activity or by the workings of a gasoline engine. The anxiety and fears

which beset us all are not removed by a knowledge of the
higher criticism of the Bible. Something more fundamental
than creeds gives zest to the hearts of believers. With its
adventurous flair for heroism, its out-thrust towards some
unifying mystery and unchanging reality greater than
itself, its welcome of values outdistancing the reach of
mere intellect, life is more than any of its elements. And
it is in its capacity to ennoble and forward human personality rather than in a knowledge of origins that the real
test of any world-view lies. From such a test "naturalism," with its denial of personal worth, apparently shrinks;
to such a test religion gladly yields.

2. Though faith be an adventure refusing to be held in
check by cautious logic, it is none the less reasonable. For
its attitudes are not without rational defense. Its world
view is tenable in the light of all the facts "naturalism"
can muster and of those other facts "naturalism" ignores.
Yet here discrimination is greatly needed. In much popular
theology, the religious world view is tantamount to a belief
in the literal accuracy of the Bible. From such a premise
some persons claim for it reliability as a revelation of
scientific facts. Particularly do they claim for the biblical
account of the origin of man an authority which excludes
acceptance of evolution. The activities of such teachers
have been considerable of late and the old "conflict between
religion and science" has often blazed forth in legislatures,
colleges and religious gatherings. Such opposition to
scientific teaching relative to the subject in dispute is not
strange. Scientific opinions have often been prematurely
or immaturely set forth and opposition to religious belief
has often been characterized by a dogmatism equal to that
of the theologian at his worst. When one recalls the place
of the Bible in Christianity, especially in Protestantism,
it is easy to see that an assertion of its untrustworthiness
would cause consternation and resentment. For in the

minds of many persons who form the very leaven of morality in our social life, belief in the Bible is indispensable to religion itself. Indeed, whatever be one's theological tenets, the religion of the Bible is, when properly understood, the very core of modern Christianity.

The champions of the inerrancy of the Bible in all fields of knowledge are consistent when they reject certain teachings of scientific authorities as opposed to beliefs and teachings of the biblical writers. It is certainly not consistent with various sciences to hold that the earth is flat, lying between masses of water below, around and above it; that day and night existed before there was any sun or other celestial body; that the sun could stand still; that man was made by God literally blowing into a mass of dust in human shape; that the center of emotion and thought is the belly or the kidneys; that life is in the blood; that there is a cave under the earth to which the spirits of the dead go. To the historically minded Christian, however, such discrepancies constitute nothing serious. They are seen to be the current ideas in which religious experience naturally clothed itself. The essence of the religious world-view is, therefore, not involved.

Nor is the religious world-view to be limited to one of its forms—supernaturalism. Rightly used such a term is legitimate. It can be so defined as to mean the existence of more or less perceived realities other than those disclosed by "natural" sciences, like chemistry and physics. But, unfortunately, such is not its usual meaning. In common parlance supernaturalism means not simply a transcendency of God to the world, but His disregard of causal sequences in nature. In ordinary events He is not concerned. He is thought of as a creator who maintains a general oversight of a going universe, but who does not do more than direct from a distance the operations of natural forces. Occasionally, however, He is thought of as interfering in the usual course of events for the purpose of showing not only that He exists, but that He is interested in human affairs, displaying His power in order to strengthen men's faith and lead them in the paths of righteousness. That is to say, He works miracles—events which are outside of the course of causal relations.

The extent to which this supernatural—or contranatural —activity of God is recognized varies according to ecclesiastical teaching. The Roman Catholic would hold that miracles are still wrought; most Protestants would say that the day of miracles closed with the closing of the biblical canon. In both alike, however, the attitude toward the possibility of the divine interference with the ordinary processes of nature is of primary importance. A denial of such interference is said not only to imperil belief in the inspiration of the Bible, but also belief in religion itself. According to such a position, unless one accepts the miracles connected with such characters as Elijah, Jonah, as well as Jesus, one denies the entire religious world-view, and thus aligns oneself with atheists.

The fact that people who lived two or three thousand years before the invention of the telescope believed that the sun could stand still does not weigh heavily with a person acquainted with the history of mankind and the science of astronomy. He knows too much about the development of folk-tales and the use of imagination in a religion to take such a story at its face value. He is not prepared to deny, of course, that exceptional happenings occur, nor, if religiously minded, that these may be the expression of divine will. But to the scientific mind, miracles in any strict sense of the term are unthinkable. Accounts of miracles must be subjected to historical tests, and if such events are found to be probable, they are to be considered as unclassified operations of forces in which God is immanent. For to the historically minded student the cry for miracle is one expression of faith in personal elements in the universe—a protest against materialism.

Clearly, a religion with its world-view is not unaffected by scientific knowledge. But can it be held when such knowledge is given to-day's extent? Before answering this question we should first clearly define the world-view implied by modern religion. Both religion and science must be contemporary. The organization of religious world-views is a genetic and historical process, varying with the culture which a society may have organized. It would be altogether unfair for a man of the modern world to estimate the tenability of religion as if it were nothing other than the faith of primitive man. It would be as unfair and as unscientific to estimate the value of a modern state by the study of primitive tribal organizations. One never can understand a developed organism by merely studying its origin any more than embryology can replace physiology. The value of any modern institution is to be judged on its own merits. Its origin and history are, of course, not to be neglected, but it will survive or perish according as it serves human needs. Religion, like other social institutions, must be judged not by the faith of our fathers, but by its service to us to-day.

The completed religious world-view of Christianity is probably the most developed product of centuries of human faith and thought. It is not supernaturalistic in the popular sense, but contains the following elements of belief not admitted by "naturalism":

1. God, the immanent personal reason, purpose and activity in the universe to whom all other wills must
   conform.
2. The individuality of man, distinct from and superior
   to, although derived from, lower forms of life.

3. The actuality of personal relationship of God and man through the activity of each.

4. Sin as opposition to the will of God, the ultimate personality of the universe.

5. Salvation or help to more personal life resulting from the establishment of proper relations between the human and divine personalities.

6. Life after death as conditioned by the individual's character and relations with God.

In Christianity these elements are exemplified and focussed in the experience and teachings of Jesus Christ.

Such a world-view is obviously opposed to any mechanistic or materialistic interpretation of the world. Therein lies a real issue, far more vital than any question of biblical interpretation. Scientific interpretations of nature which deny all freedom of choice and reduce the intellectual life to chemical and physical activity are not consistent with religion, and religion is not consistent with them. It is, of course, true that the materialist often finds himself unable to break with his social heritage, but religion in any real sense of the word should logically be no more in his life than is Santa Claus in the mind of the modern man. The basal issue is very simple: waiving all terms of misleading connotations, does the universe with all its forces permit one to live in accordance with the religious point of view? Is religious faith consistent with what we know about the universe?

It will be noticed that the question is not whether the facts of the universe are consistent with some religious faith already held. All religious faith contains so many ideas and beliefs embodied in the social life of a people as to be invalid as a test of scientific knowledge. Because a man believes in God is no reason for accepting or rejecting the existence of electrons. But if there are in the activity of the universe no qualities which justify belief in cosmic reason and purpose, religion will have to go.

Many would divorce knowledge and values. Certainly it is one thing to think of power and another to think of goodness; one thing to know and another to act in accordance with knowledge; one thing to estimate a forest as potential wood pulp and another to respond to its beauty. But to recognize the difference between the world of science and the world of religion is not to affirm that the one is contradictory to the other. A man has to be religious in the universe in which he is. He cannot select from it a little universe of his own. Religious faith is either consistent or inconsistent with reality. As our knowledge of reality grows, our religion will grow more or less tenable.

To be of real social significance a religion must be as intelligent as its possessor. Men have been temporarily able to hold their religious knowledge and religious faith independent of each other, as it were in separate compartments. But in the long run, such an attempt is not successful. Sooner or later, by a sort of osmotic pressure, one belief has permeated the other.

Let us, then, put the issue quite frankly. We have inherited religion as an element of our social life. Granting that the knowledge of our universe and of life is as the scientists have described it, is religion legitimate? Is it as an expression of life itself, rationally consistent with such knowledge of reality as we have? The answer is clearly one of fact. It is no new question. There have always been those who have raised it. And in every case the affirmative answer would seem to depend upon the same grounds. Men have found in the world they knew the justification for holding and often expanding the faith which they inherited. Their faith has been legitimatized by whatever scientific knowledge was available. True, this process has often required time. New knowledge has raised new questions. When the world of reality has changed, as, for example, when scientific conviction that the world was the center of the universe was replaced by the Copernican system, there has been friction in religion at the point where the two scientific views touched each other. The mental serenity which has developed from harmony between religious faith and the world of knowledge is always disturbed by the intrusion of new facts and new generalizations. From the day of Lucretius to our own, because of intellectual honesty, pioneers in the realm of science have not infrequently been skeptics or deniers in the realm of faith. Yet their unbelief has been due less to their scientific achievements than to some inherited philosophy. Their successors in the field of culture have usually not shared their inner struggles. Either they have regarded religion as incompatible with the world of scientific realities or they have embodied the new knowledge in their religious faith.

In our own day we are experiencing this sense of friction. Churchmen have been concerned lest faith in God should dissolve before the discoveries of the various sciences. The anxiety is natural but unfounded. To-day, as in the past, the religious impulse and attitude are being enriched by our new knowledge. Formulas, it is true, have been abandoned, but new formulas enriched by new knowledge are ever appearing.

The plan of the present volume is to consider the legitimacy of the religious life and world-view. In Part I, recognized scientific authorities state briefly the chief elements of our knowledge in various fields of investigation. In Part II, other authorities describe typical experiments in furthering human welfare by proper adjustment of life to the impersonal forces in the environment of the universe; and, in Part III, religion as an essentially similar adjustment of life (but through personal relations) with our environment is found to be warranted by our expanding knowledge or reality.

The center of our thought is, therefore, not philosophical or metaphysical, but practical. We do not seek at the outset to prove the existence of God. Religion as a phase of social life in the universe of realities is to be judged. Despite our admitted inability to reach knowledge of ultimate reality we raise this question: In the light of scientific method and fact, has the most developed religion a right to our confidence in any degree comparable with the claims, let us say, of agriculture and preventive medicine? And if such confidence is justified, what contribution does science make to the religious life?

PART I

THE WORLD GIVEN BY SCIENCE

CHAPTER II

THE METHOD OF SCIENCE
By WILLIAM E. RITTER [1]

IT is fortunate that the subject of the opening chapter of this Part is worded so as to require its writer to discuss the procedure by which science reaches truth rather than what science holds to be the nature of truth. Nevertheless, it should be said that science is not wholly devoid of at least an opinion as to what truth is. Otherwise it might be inferred that when a scientist begins an investigation he is in utter darkness as to what he is doing. This, of course, is not the case. What is dark to him is the particular thing to be investigated; not everything round about it. Some things, indeed many things, are already in the light. He ever seeks not all light (or truth), for some of it he already has; but more light.

In relation to the general purpose of the book, that of showing how science helps religion, this question of the scientific conception of the nature and accessibility of truth is fundamental. In all the history of human thinking, there have been three main roads for the would-be truth-seeker: the road of authority, or testimony of tradition and revelation; the road of the mystic, or intuitive perception; and the road of the scientist, or the testimony of sensory experience. There has been a tendency to claim that these various roads lead to different goals; that certain kinds of truth are not accessible to the scientist, but

[1] Former Director Scripps Institution for Biological Research, University of California.

only to the mystic directly, or to others indirectly through authoritative revelation. If there be any such bounds to the kingdom of knowledge, science has no means of recognizing them. No man can legitimately excuse himself from honest thinking in any field of human experience, where material for observation and deduction are available. Nor have we any reason to fear that honest and courageous thinking will be less fruitful in the fields of religious experience and human relationships than it has been in dealing with physical and chemical phenomena. Seeing God in the universe is no more difficult than seeing electrons there. We have ample testimony concerning both; concerning neither have we complete and final knowledge.

"Howbeit when he, the Spirit of truth, is come, he will guide you into all truth" (John xvi: 13).

Never can I forget an incident connected with this familiar passage which occurred when I was a mere stripling in both science and religion, very eager in the former and rather faltering in the latter. It was at the mid-week prayer meeting of the church to which I belonged. I ventured to raise the question as to whether the "all truth" of the passage might be taken to mean what it says. Whether it might include, for example, the truths of chemistry in which I then happened to be specially interested. What has clung most tenaciously to my mind was the decisive, and almost annoyed, way in which the minister who was presiding informed me that the passage had nothing to do with that kind of truth. "It refers," he said, "only to sacred truth." I may mention, by the way, that no small part of my recollection of this incident has been the wonder I have always felt as to whether this minister's later loss of religious faith, and his simultaneous weakening in moral character, had anything to do with his notion of truth. His hold on the realities of life might have been stronger had he realized that truth is one, that it cannot be divided into sacred, chemical and physical varieties in such a completely isolable way as he supposed.

The method, then, by which science reaches truth must have much in common with the method by which any knowledge whatever, indeed all knowledge, reaches truth.

Science, it has been said, in slightly varied wording thousands of times during the last century, is only common knowledge refined and extended. But despite the innumerable times this has been approvingly repeated, its truth seems not to have sunk deeply into the minds either of intelligent people generally or of many scientists and philosophers.

If science is really an outgrowth of common knowledge, then the right starting place for a study of scientific method is a study of the method of common knowledge. Now, the beginning of common knowledge is inseparably connected with common, sensory experience. And the beginning of such experience is in turn inseparably connected with the beginning of independent life in the individual.

The genetic course of science seen in broad outline is clear: scientific knowledge (or science) develops out of common knowledge; common knowledge develops out of common experience, and common experience develops out of life itself as a part, or inescapable by-product, of the very function of living. Consequently, a study of the method of science would have to cover this whole series: method of scientific knowledge, method of common knowledge, method of sensory experience, method of life. Nothing short of such an expanse of the study would do, no matter how little or how much time and effort were put upon it.

While recognizing that our study, in order to be sound, must cover this entire series, we may begin the study at either end of the series or anywhere in it. As a matter of convenience (or possibly of predilection due to the author's long-established practices as a naturalist) we will begin at the lower, more general, end of the series.

COMMON KNOWLEDGE THE BEGINNING OF SCIENCE

The method of human life, at its very foundation, what is it? Thanks to a vast body of common information on this subject, backed up by a very large body of technical information, we are able to answer a determinative part of the question with an assurance that has no trace of doubt. The very first thing in the method of life of every independent human being is the first breath it takes as an infant just born into the world. Or, stating the matter a trifle more in conformity with the idea of method, the first act of independent life is a response to the complex of stimuli from the world the infant has just entered, resulting in a flow into its lungs of air containing its peculiarly life-giving element, oxygen. It is one of the commonest items in common place knowledge that unless this first act of breathing is performed the independent life does not begin, and further that the dependent life which is brought over from the mother can endure for only a few hours at most. Furthermore, the method of life includes a continuance of the act of breathing to the very end.

The second imperative thing in the method of life is the act of food-taking. Though eating is less immediate in its imperiousness than is breathing, it is not a whit less absolute in the long run.

What a world of confused and futile thinking has resulted from ignoring these indubitable facts of every human life! It is no perversion of the truth to say that many persons, speaking not alone on the plane of common knowledge, but as well on the planes of philosophy and religion and even of science, talk as though being born and living as infants and small children had somehow never been necessary in their cases.

So much as an indispensable look at the first member, the method of life, of the series we are examining.

Now, notice how this look carries us forward to the second member of the series, sensory experience. The act of breathing consists, as stated, essentially in the response of the infant's respiratory mechanism to the stimulating influence of the air and other things with which its little body has never before come in contact. And the acts with which food-taking is initiated and continued are even more definitely responses to external stimuli. The quickness and vigor with which the hitherto but slightly mobile mouth begins its characteristic activities known as sucking the moment the lips touch in proper fashion almost any object are traits so peculiarly infantile that reference to them in a discussion of truth-seeking by science may strike some readers as far fetched if not puerile. But their consideration here is particularly important because in connection with them an exceedingly important new element enters into the series of steps by which knowledge is obtained. The element referred to is the power of discrimination. One of the earliest and sharpest manifestations of this power in the individual is the recognition that sucking certain objects which can be taken into the mouth will yield food while doing the same thing to others will not. Surely nobody needs reminding of the enormous rôle the power of discrimination relative to food has played in the history of mankind.

The new-born infant then breathes and takes food as a response to various internal and external stimuli. If he does not so respond he dies. But in the very act of making these essential responses, he accumulates a relatively great store of experiences, upon which he at once exercises the power of discrimination. Getting knowledge, and knowledge trustworthy enough to order his life by, has begun with the first breath and the first meal.

The question has been much discussed in speculative philosophy of how it is possible for responses to the stimulations produced by external objects to become transformed into, or in any way enter into, our conscious knowledge of these objects. Indeed so great seem to be the difficulties of this question when they are viewed as speculative philosophy has been wont to view things that some philosophers have doubted the possibility of any such transformation. They have doubted or even denied that the external world is really intelligible after all. They have contended that our sensory impressions do not help us to know fundamental truth, but are illusory.

What we should have to say to such a theory as this would seem to be clearly indicated by what has just been said about knowledge and life: The proof for us that the world is intelligible is the fact that we can and do *live in it.* For us the problem of how we are able to know the world is part and parcel of the problem of how we are able to live in and by means of the world.

The entrance into the series of this power of discrimination marks the passage from the method of sensory experience, pure and simple, to, or at least toward, that of the next member of the series, namely common knowledge. For whatever else knowledge of any kind may be, surely some power to discriminate is part of it. Recognition of the power thus far down in the series is particularly relevant to our undertaking for this reason: Its entrance marks not only the transition from mere sensory experience to common knowledge, but also the beginning of that improvement in the method of knowledge-getting which, going on, results at least in the refinement and extension of common knowledge accepted at the outset of our discussion as a true characterization of science.

The method of science might be characterized as the method of *greatly improved and always improving common knowledge.* This is really only another way of saying that the method of science consists in, or at least involves, the constant improvement of observation. See not only well but ever better; hear not only well but ever better; touch not only well but ever better; smell not only well but ever better. The carrying out of these injunctions are absolutely indispensable to the method of science. Nor need one be in the least quandary as to how well the observations should be made in any particular case. Well means well enough to remove the last trace of doubt as to whether the thing observed is what it is supposed to be. And this, of course, makes clear how much, if any, improvement of observation in any given instance is requisite.

While observation upon, or experience with, many of the commonest things in our environment run on with great ease and trustworthiness, as to many others, much of correction and refinement of observation are necessary. As an example of such easy-running experience consider much of that connected with the sun. Rarely, indeed, does one have to "take a second thought" as to the identity of this object. There is little chance under ordinary circumstances of mistaking it for anything else, or mistaking anything else for it.

And much the same may be said of the moon, though not quite the same. One is more apt to hesitate as to whether a particular light on the horizon is the rising moon or a fire than he is as to whether it is the rising sun. In other words, one has to correct his observation more frequently for attaining certainty in the case of the moon than in that of the sun.

On the other hand, if one wants to know "where lies the wind" at a time when there is little of it, he may find it necessary to resort to various expedients of observation before he reaches certainty. Turning first in one direction and then in another so as to get the effect of all there is of it full in the face may suffice. But again it may not. It may be necessary to drop some very light object, as a feather, and notice which way it veers as it falls. And there are other more or less familiar methods that may have to be resorted to before the question is fully settled to the satisfaction of the inquirer.

These humble examples are enough to illustrate the fact that refinement of method in observation begins very early and is apt to be necessary at any time for the attainment of certainty, no matter whether the knowledge sought be called scientific or common.

VERIFICATION AND CORRECTION OF OBSERVATION

There are two things in connection with improving the methods of observation that are so important we must give them special attention. One of these is what is known as verification; the other is the correction of sensory illusions.

We may be almost, but not quite, certain that, for example, the light on the horizon is due to the moon and not to a fire. In order to make "assurance doubly sure," we may go to a more favorable place for observing; or we may have to wait awhile and watch developments. But sooner or later, if all goes well, our efforts are successful. Our supposition is confirmed—or the contrary—as the case may be. We have verified, or the contrary, our first impression or belief. We have proved, or disproved, our guess or supposition, or, in the more formal language of science, our hypothesis.

Obvious, is it not, that this sort of thing we do every day of our normal lives? And no one who has had laboratory instruction in even the rudiments of any natural science can fail to recognize that the same sort of thing was no small part of the method he there followed. Every boy and girl begins to be a *formal* verifier in exactly this sense the moment he or she begins the practical study of any natural science whatever; as indeed they have been *informal* verifiers all their lives.

Nobody escapes the experience of being sometimes deceived as to just what it is that his senses tell him. Indeed such deception is so apt to occur and to lead us so far astray that it is made much of in support of the widely held theory that intellect, reason, intuition or some power of the mind wholly independent of sense perception is the only really essential thing in attaining truth.

This theory ignores the very evident fact that, while our senses lead us astray, it is only through our senses that we know we *are* astray, or ever find our way back. For instance, when a straight stick is half immersed, slant-wise, in a body of clear water, it appears to have an angle in it just at the surface of the water. This is often used to illustrate the untrustworthiness of sensory knowledge. But how do we discover and then correct the illusion in this case? Does our reason or our intuition or some other higher power of our mind, acting entirely alone, tell us we are wrong and set us right? Not at all. The illusion is discovered and corrected by bringing still more sense experience to bear on it. We take the stick out of the water or change the direction of its penetration into the water as the most direct way of coming at the real truth of the matter.

There are illusions much more far-reaching and difficult of discovery and rectification, such as that of the apparent rising and setting of the sun and other heavenly bodies. How was the universal error of sense experience here mentioned discovered and rectified? The history of modern astronomy beginning with the innumerable observations *and later* the calculations of Copernicus, and the telescopic observations of Galileo is too familiar a story to need any retelling here.

No one who has the slightest acquaintance with the history of man's knowledge of the universe can fail to recognize the great part played by these two ways of improving the methods of getting knowledge, namely, verification and correcting illusions. The refinement of methods of observation which has contributed to the conversion of common knowledge into science has been and continues to be enormously important. But this is not the whole story: extension as well as refinement of method of observation is essential to convert common knowledge into science.

The Rôle of Curiosity

From what we have seen of the refining operations, it would be justifiable to say that these arise from man's inherent *caution*. Quite similarly it may be said that the extending operations arise from his inherent *curiosity*.

That the attribute of being curious about things is enormously developed in man, taking him all in all, is obvious enough. Innumerable of the discoveries, small and great, that have made up so large a part of his career upon the earth, are proof positive of his natural eagerness to know ever more and more about the world around him, and that without much apparent reference to what use the things known will be to him. Taking him on the whole, as a species, man just wants to know and is determined to know.

Desire to push on and ever on in the attainment of knowledge—of truth—is as deep-seated in man's nature as is his ability to have sensory experience, and for exactly the same basic reason, *viz.*, the inseparability, finally, of stimulus from response in the "stimulus-response polarization."

Now, this attribute of curiosity, with its concomitant desire to push on, is something very remarkable, something very precious. Indeed, so remarkable and so precious is it as a personal possession that those who have never come close enough, in both body and soul, to the heart of nature to discover the universality of what is remarkable and precious there also are wont to believe this possession to be something utterly unique, something wholly apart from nature, something, in other words, mysterious. The fruits of failure to push on in the pursuit of natural truth are pretty sure to be somewhat alien from nature, something of distrust in fellow man, something of either morbid or ecstatic loneliness of life. But it is not the fruits of failure to forge ahead that primarily concern us now. Rather is it the method of such forging.

THE HYPOTHESIS

The most peculiar and important of all the instruments of this on-pushing in pursuit of truth is the hypothesis. If one were obliged to specify some one thing as more characteristic than any other of what the mind does whenever it moves forward or seriously tries to in response to its everlasting urge for more information, more knowledge, more truth, I think he would have to choose hypothesis-making and hypothesis-proving (or -disproving) as the one. To the hypothesis we are ever appealing.

Nor is this really any more characteristic of scientific knowledge-getting than of common knowledge-getting; for as a matter of fact every time an intelligible question is asked an hypothesis is implied. This is so for the simple reason that what is uttered is a question, is something not known but desired to be known. But the fact that knowledge about the unknown thing is *desired* is conclusive proof that it is not wholly outside the pale of total knowledge. Things about which we are absolutely ignorant, about which we know nothing either as to the thing itself or anything pertaining to it, we are absolutely incapable of asking any intelligible question about.

But again the fact that we are able to ask an intelligible question about an unknown thing implies that we are able to give a *provisional,* that is, a *possibly true* answer to it. Now, this possibly true answer is what has received the name hypothesis. And it is important to notice that in almost every case (probably, indeed, in every case) where a hypothetically true statement can be made about an unknown thing, at least one other such statement can be made about the same thing. In other words, more than one hypothesis is possible for practically every problem. The very essence of a question implies that in the mind of the questioner at least more than one answer is possible.

Glance now at a list of questions taken at random from our modern life.

Is it raining? Is the water cold? Is dinner ready? What is the cause of cancer? How many eggs may a white leghorn hen be reasonably expected to lay in a season? What is the population of China? Do the straight-ahead rays from the headlight of an automobile running at forty miles an hour travel any faster than those directed straight behind from the tail-light? Is democracy a success? Is lead heavier or lighter than mercury? What animal species was the ancestor of the human species? How many square rods of land are there within the horseshoe bend of Pebble Brook on South Slope Farm?

Inspect any such group of questions and you notice this about them: You find it quite impossible to divide them up into two well-defined lots, one of which falls into the realm of common knowledge and the other into that of science. A few of them, like: Is dinner ready? Is it raining? would probably be allowed to pass as belonging strictly to the realm of common knowledge; though even here some one might suggest that the answer would depend on *where* the raining question applies, and *whose* dinner is intended. On the other hand a few of them, like that about light rays and that about the cause of cancer, would probably be assigned without hesitation to the realm of science.

But most of them clearly fall into both realms, depending on how full the answer is to be, and what purpose it is to serve.

But important as is the fact that questions and answers, problems and their hypotheses, grade insensibly from common into scientific knowledge—overlap in the most intimate way as between these two kinds of knowledge—a still more important thing about hypotheses must be noticed. This concerns the kind of answers which can be given to the questions, or what amounts to the same thing, the kind of proofs to which hypotheses are susceptible.

To illustrate : such a question as, Is it raining at some particular place ? can usually be answered with *absolute certainty*. It either is or it is not. And common knowledge is every whit as competent to answer as is scientific knowledge, *ordinarily*. Under certain circumstances it may not be. Such a question, too, as, Is lead heavier or lighter than mercury ? can be answered with *absolute certainty*. But common knowledge would be less trustworthy than scientific knowledge in this case.

The point is that a great array of phenomena in both nature and art is susceptible of being interrogated, hypothetically answered, and the answers made positive and final by sufficient effort.

The inevitable conclusion appears to be that there is an extremely important sense in which both common and scientific knowledge can reach ultimate truth. The psychobiological foundation on which rest the logical principles of identity and difference justifies this. Every object is absolutely itself, and not any other object. No rock or insect or chair is any other rock or insect or chair, or any other object whatever. If there is any thing absolute about any kind of knowledge it would seem to be knowledge of the kind here illustrated.

But while there is a great class of questions concerning the objective world to which absolutely certain answers can be given, there is another and perhaps greater class to which such answers cannot be given, either by common or by scientific knowledge. Consider such a question as, Is the water cold? "Cold" is absence of "heat," and "heat" is, in turn, dependent upon molecular activity of the water which activity can be anything between the point at which water becomes gaseous at one extreme and ice at the other. Manifestly, therefore, there can be no once-for-all absolute "yes" or "no" to the question. But it is important to notice that for all such practical purposes as surf bathing and most domestic and many laboratory uses, answers can readily be given that are true enough to satisfy the particular needs.

But what can we say about such a question as that concerning the ancestor of the human species? This is what we can say: having marshaled every bit of knowledge we have concerning the nature and development of man on the one hand, and concerning the nature and development of brute animals on the other; and having thought as fully and carefully as possible about the bearing of all this knowledge on the question, the conclusion we reach is that such-or-such (for our purpose here never mind what) animal species is *more probably* the ancestor of man than is any other species. The most probable truth is the nearest we ever get or ever can get to absolute truth in any of the problems of this vast class.

This matter is vital and we must reach the greatest possible clarity upon it. We can do this by noticing again the close kindred of common knowledge and scientific knowledge. The entire safety and hence great readiness with which we accept probable truth in numberless of the commonest affairs of life is the point to be made.

Take this very matter of origination among human beings. In the room with me at the moment I write these words are three persons, all entire strangers. Until now I have never seen any of them or even heard of them. I ask myself, Were they all "born of woman"? Were they all once new-born infants? Without a long siege of question-asking, record-searching, and affidavit-taking, what answer can I give? The *probability* that each of these persons was born in the usual way, was once an infant, is so great that for all ordinary purposes a question about it would not occur to any one. That, however, such questioning does arise is illustrated by the number of personages that have been accounted of divine or superhuman origin in the history of mythology and various religions. And question as to the particular parentage of such and such a person is, of course, frequently legitimate and necessary. When, for example, a government would issue a passport to one of its citizens, it cannot afford to rest on the probability as to the citizen's parentage. It must have evidence; and what can be accepted as sufficient evidence is highly significant. A probate court must be still more exacting of evidence as to parentage where inheritance of property is involved.

"Probable" and "Absolute" Truth

The main point illustrated is the implicit trust we often put in *probable* as contrasted with *absolute* truth. Look around you at any moment and ply yourself with questions relative to the innumerable objects that go trooping through your field of vision. You readily notice what a vast number of your questions you are able to answer offhand, with entire satisfaction on the basis of what is in the highest degree *probable*. The gravel bank by the roadside was probably deposited by ice but possibly by water. The oak tree in the field was almost, but *not quite*, certainly produced from an acorn. In very many cases, like the question of parenthood just discussed, probability may be elevated to certainty. In many other cases this is not possible. We are obliged to do the best we can on the basis of so much truth, and of such probability of truth, as can be attained.

Our question about the ancestry of the human species falls into this class. Also the question, Is democracy a success? comes here as do very many of the personal and social questions most vital to human welfare.

It is, consequently, with reference to such questions that human wisdom gets its severest tests both in coming to mental decisions and in acting on them.

In no connection does the constancy and orderliness of the universe contribute more to human life and human welfare than in connection with this very matter of probable truth. The astronomer predicts the eclipse, the farmer plants his seed, the mariner goes to sea, the manufacturer plans his operations for the coming year, the school makes its budget, all on the *probability* that things will go on in the immediate future much as they have gone in the past. And the degree of probability as to outcome in every case is dependent upon the extent and accuracy of knowledge (past experience) which can be brought to bear upon the problem.

And here an exceedingly important methodological element comes into the effort to answer questions, especially if they are of considerable complexity. That element involves the fact that all phenomena present themselves in some quantity. There is always some number of objects involved, and each of these has something in the way of size, weight, rate of movement, density and so forth. In other words, problems and hypotheses can always be treated mathematically to some extent as well as in the coarsely descriptive fashion most familiar to us.

The farmer knows from experience about the *time* of year each crop should be planted; he knows about how *much* seed to plant to each unit of his field; he knows about how *many days* will be required to mature the crops, etc., etc.

When it comes to such problems as those of the recurrence of eclipses, the countings and measurings required are so relatively few and simple that a wonderfully high degree of probable truth can be reached as to what will take place at a future time.

But the fact should never be lost sight of that a *calculated* occurrence as of an eclipse is an hypothesis and is never proved absolutely until the occurrence itself has been observed. And it is further important to note that wonderfully close as the astronomers are accustomed to make their predictions, rarely, if ever, I am told, does it happen that the predictions and the observations exactly correspond, though the deviations may be negligible for the purpose of observing the eclipse.

The only absolute truth there is connected with proving astronomical or any other hypotheses in ''exact science'' is that while the predicted occurrence is under observation the certainty of it is absolute. Even the best records of it pass immediately into the vast realm of probable truth with the passage of the occurrence itself, so far as future reference is concerned.

So great are the natural powers for quantitative calculation possessed by some persons (mathematicians) and so vast has become the body of knowledge (mathematics) built up through these powers, that it is not surprising that some of these persons should fail to recognize the genetic relation of their power and their knowledge to the sensory experiences, individual and racial, of mankind as a whole.

A considerable number of mathematicians feel certain that their minds have access to truth in their particular realms which is wholly independent of, and far superior to, sensory experience of any kind. This appears to be what they mean by their intuitions and by the supposed *a priori* character of the basic postulates of mathematics.

But those who conceive mathematics in this way have never, so far as I know, fully reconciled the conception with the historic beginning of arithmetic in counting the common objects of all mankind's daily affairs, and of the origin of geometry in measuring land and other areas.

But a still more serious difficulty, as it seems to me, is encountered by the intuitional theory of mathematics, in that identification of all knowledge with individual life which we saw early in this discussion.

THE FAITH OF METHODOLOGY

The conclusion that sensory experience plays a part even though, as in pure mathematics, a seemingly small and remote part, in the method by which science reaches truth is supremely important for the purpose of this book, namely of showing how science helps religion. This is so because it prepares the way for the unification not only of all Truth but of all Faith.

What more solid ground, what better tested ground, for faith can anybody point to than that which rests on the four pillars at which we have glanced in the preceding pages? Let us mention these seriatim: (1) The life-creative and life-sustentative power of nature. (2) The knowability of at least vast provinces of nature. (3) The forecasting power of the mind into unknown provinces of nature. (4) The constancy and orderliness of nature making possible implicit trust in the probability of truth as touching numberless of the most vital concerns of human life.

When men shall have made really their own the facts and seemingly unescapable conclusions, sketched in the preceding pages, they will have to recognize the enormous part played by faith in all knowledge, scientific as well as common.

KNOWLEDGE, FEELING AND LIFE

It remains to conclude this discussion with a brief reference to an aspect of truth and truth-seeking not yet touched.

The reader will remember that use was made in opening the discussion of a familiar New Testament passage concerning the nature of truth.

For closing the discussion we will make use of another classical utterance on the subject: "They who know the truth are not equal to them that love it." One of the particularly striking things about this saying of Confucius is its modernness. It is accordant with recent discoveries in psychology and physiology. Although it sets knowing and loving in a measure of opposition to each other it also virtually affirms that both are at their best when they are joined together. The inference is clear that while one may know or seem to know truth without loving it, he ought not so to do. But since love has its headquarters in the emotional side of our nature while knowledge belongs peculiarly to the rational side, it would follow from this Confucian doctrine that intellect and emotion ought to go hand in hand in the pursuit of truth. And this is exactly what present-day psychobiology justifies. The evidence is now conclusive from technical knowledge, that no such isolation of elements in our spiritual life occurs or is possible as has been taught by certain systems of philosophy and metaphysics, some of which have had, and still have, great vogue. Sensation, feeling and emotion are not apart from thought, reason and intellect in any such way as a great variety of theories of Mind as opposed to Body would have us believe. There is no such thing as Pure Reason, all-powerful and infinitely noble, but caught for a time as by some strangely fatal accident, in a poor, weak, groveling body. It is exactly in its demonstration of connectedness between mental processes and bodily processes and between different bodily processes that psychobiology has lately scored its greatest triumphs.

From what has been said it is recognizable that the familiar characterization of science as emotionless and cold, as a thing of the head and not at all of the heart, must be false. Such a conception of science is about as much of a travesty on its real nature as could be invented.

Life, Knowledge, Truth—three in one and one irrefragably composed of three—that is where we are led by close scrutiny of the method by which science reaches truth.

Nor are the life, the knowledge, the truth thus before us wholly at variance with the meanings we attach to those words as we use them in our daily lives, our common tasks. However large may be the life implied, included in it are all our myriad experiences as beings of flesh and blood; of head and heart and hand; of feelings, emotion and passion; of reason, thought and intelligence.

CHAPTER III

THE STRUCTURE OF MATTER

By ROBERT A. MILLIKAN [1]

WITHIN the last three decades there has been
discovered beneath the nineteenth-century world
of molecules and atoms a wholly new world
of electrons. The properties of these electrons have
been carefully studied and they have been found to
be of two kinds, negative and positive, which are,
however, exactly alike in the quantity of electricity
existing in each, that is, in charge, but wholly different in
inertia or mass, the negative electron being associated with
a mass which is but 1/1845 of that of the lightest known
atom, that of hydrogen, while the positive appears never
to be associated with a mass appreciably smaller than that
of the hydrogen atom. Indeed we think it is the hydrogen
atom itself when its single and very light negative electron
has been removed from it. The purpose of the present
paper is to discuss what the physicist, as he has peered with
his newly discovered agencies—cathode-rays, x-rays, radioactivity, ultra-violet light, etc.[2]—into the insides of atoms,
has been able to discover regarding their structure, and to
show how far he has gone in answering the fundamental
question as to whether or not the electrons are the sole

---

[1] Professor of Physics, California Institute of Technology.

[2] For a historical review of the discovery of the electron and a
much more complete treatment of the subject of this chapter, the
reader is referred to the author's *The Electron*, etc., published
by the University of Chicago Press, 1917, revised, 1924.

building stones of the atoms, and if so how they are arranged and what they are doing within the atom.

THE SIZES OF ATOMS

The isolation and exact measurement of the primordial electrical unit, the electron, gave us at once a knowledge of the exact number of molecules in a cubic centimeter of a gas. Indeed, we can now count the exact number of molecules in any given volume or in any known weight of any homogeneous substance with even more certainty than we can count the population of a city or a state. Thus, the molecular population of a cubic centimeter of ordinary air is exactly 27.05 billions of billions.

But although the exact absolute weights and numbers of atoms have only recently become known through the precise measurement of the electronic charge, we have for years had satisfactory evidence as to the relative weights and the relative sizes of different atoms and molecules. For we have known for a hundred years, at least, that different gases when at the same temperature and pressure possess the same number of molecules per cubic centimeter (Avogadro's rule). From this it is at once evident that, as the molecules of gases eternally dart hither and thither and ricochet against one another and the walls of the containing vessel, the average distance through which one of them will go between collisions with its neighbors in a gas at a given pressure will depend upon how big it is. The larger the diameter the less will be the mean distance between collisions—a quantity which is technically called "the mean free path." Indeed, it is not difficult to see that in different gases the mean free path l is an inverse measure of the molecular cross-section. The exact relation is

$$l = \frac{1}{\pi n d^2 \sqrt{2}}$$

in which d is the molecular diameter and n is the number of molecules per cubic centimeter of the gas. Now, we have long had methods of measuring l, for it is upon this that the coefficient of viscosity of a gas largely depends. When, therefore, we have measured the viscosities of different gases we can compute the corresponding l's and then from equation (1) the relative diameters d, since n is the same for all gases at the same temperature and pressure. But the absolute value of d can be found only after the absolute value of n is known. If we insert in the foregoing equation the value of n mentioned above, it is found that the average diameter of the atom of the monatomic (one atom) gas helium is 2×10^{-8} cm. (two hundred millionths of a centimeter), that of the diatomic (two atoms) hydrogen molecule is a trifle more, while the diameters of the molecules of the diatomic gases, oxygen and nitrogen, are fifty per cent larger. This would make the diameter of a single atom of hydrogen a trifle smaller, and that of a single atom of oxygen or nitrogen a trifle larger than that of helium. *By the average molecular diameter we mean the average distance to which the centers of two molecules approach one another in such impacts as are continually occurring in connection with the motions of thermal agitation of gas molecules—this and nothing more.*

As will presently appear, the reason that two molecules thus rebound from one another when in their motion of thermal agitation their centers of gravity approach to a distance of about 2×10^{-8} cm. is presumably that the atom is a system with negative electrons in its outer regions. When these negative electrons in two different systems which are coming into collision approach to about this distance, the repulsions between these similarly charged bodies begin to be felt, although at a distance the atoms are forceless. With decreasing distance this repulsion increases very rapidly until it becomes so great as to over-come the inertias of the systems and drive them asunder.

THE RADIUS OF THE ELECTRON FROM THE ELECTRO-MAGNETIC THEORY OF THE ORIGIN OF MASS

The first estimates of the volume occupied by a single one of the electronic constituents of an atom were obtained from the electro-magnetic theory of the origin of mass, and were, therefore, to a pretty large degree speculative, but since these estimates are strikingly in accord with results which follow from direct experiments and are independent of any theory, and since, further, they are of extraordinary philosophic, as well as historic, interest, they will briefly be presented here.

Since Rowland proved that an electrically-charged body in motion is an electrical current the magnitude of which is proportional to the speed of motion of the charge, and since an electric current, by virtue of the property called its self-induction, opposes any attempt to increase or diminish its magnitude, it is clear than an electrical charge, as such, possesses the property of inertia. But inertia is the only invariable property of matter. It is the quantitative measure of matter, and matter quantitatively considered is called *mass*. It is clear, then, theoretically, that an electrically charged body such as a pith ball must possess more mass than the same pith ball when uncharged. But when we compute how much the mass of a pith ball is increased by any charge which we can actually get it to hold, we find that the increase is so extraordinarily minute as to be hopelessly beyond the possibility of experimental detection. However, the method of making this computation, which was first pointed out by Sir J. J. Thomson in 1881, is of unquestioned validity, so that we may feel quite sure of the correctness of the result. Further, when we combine the discovery that an electric charge possesses the distinguishing property of matter, namely, inertia, with the discovery that all electric charges are built up out of electrical specks all alike in charge (electrons), we have made it entirely legitimate to consider an electric current as *the passage of a definite, material, granular substance along the conductor*. In other words, the two entities, electricity and matter, which the nineteenth century tried to keep distinct, have lately begun to look like different aspects of one and the same thing.

But, though we have thus justified the statement that electricity is material, have we any evidence as yet that all matter is electrical—that is, that all inertia is of the same origin as that of an electrical charge? The answer is that we have *evidence,* but as yet no *proof.* The theory that this is the case is still a speculation, but one which rests upon certain very significant facts. These facts are as follows:

If a pith ball is spherical and of radius a, then the mass m due to a charge e spread uniformly over its surface is given, as was first shown by J. J. Thomson, by the equation (2):

$$m = \frac{2}{3} \frac{e^2}{a}$$

The point of special interest in this result is that the mass is inversely proportional to the radius, so that the smaller the sphere upon which we can condense a given charge e the larger the mass of that charge. If, then, we had any means of measuring the minute increase in mass of a pith ball when we charge it electrically with a known quantity of electricity e, we could compute from equation (2) the size of this pith ball, even if we could not see it or measure it in any other way. This is much the sort of a position in which we find ourselves with respect to the negative electron. We can measure its mass, and it is found to be accurately 1/1845 of that of the hydrogen atom. We have measured accurately its charge and hence can compute the radius a of the equivalent sphere, that is, the sphere over which e would have to be uniformly distributed to have the observed mass, provided we assume that the observed mass of the electron is all due to its charge.

The justification for such an assumption is of two kinds. First, since we have found that electrons are constituents of all atoms and that mass is a property of an electrical charge, it is, of course, in the interests of simplicity to assume that all the mass of an atom is due to its contained electrical charges, rather than that there are two wholly different kinds of mass, one of electrical origin and the other of some other sort of an origin. Secondly, if the mass of a negative electron is all of electrical origin, then we can show from electro-magnetic theory that this mass ought to be independent of the speed with which the electron may chance to be moving unless that speed approaches close to the speed of light. But from one-tenth the speed of light up to that speed the mass ought to vary with speed in a definitely predictable way.

Now, it is a piece of rare good fortune for the testing of this theory that radium actually does eject negative electrons with speeds which can be accurately measured and which do vary from three-tenths up to ninety-eight hundredths of that of light. *It is further one of the capital discoveries of the twentieth century that within these limits the observed rate of variation of the mass of the negative electron with speed agrees accurately with the rate of variation computed on the assumption that this mass is all of electrical origin.* Such is the experimental argument—not completely adequate since the development of the theory of relativity—for the electrical origin of mass. From it we can proceed.

Solving, then, equation (2) for a, we find that the radius of the sphere over which the charge e of the negative electron would have to be distributed to have the observed mass is but 2×10^{-13} cm., or but 1/50,000 of the radius of the atom (10^{-8} cm.). From this point of view, then, the negative electron represents a charge of electricity which is condensed into an exceedingly minute volume. In fact, its radius should be no larger in comparison with the radius of the atom than is the radius of the earth in comparison with the radius of her orbit about the sun.

In the case of the positive electron there is no direct experimental justification for the assumption that the mass is also wholly of electrical origin, for we cannot impart to the positive electrons speeds which approach the speed of light, nor have we as yet found in nature any of them which are endowed with speeds greater than about one tenth that of light. But in view of the experimental results obtained with the negative electron, the carrying over of the same assumption to the positive electron is at least natural. Further, if this step be taken, it is clear from equation (2), since m for the positive is nearly two thousand times larger than m for the negative, that a for the positive can be only 1/2,000 of what it is for the negative. In other words, the size of the positive electron would be to the size of the negative as a sphere having a two-mile radius would be to the size of the earth. From the standpoint, then, of the electromagnetic theory of the origin of mass, the dimensions of the negative and positive constituents of atoms in comparison with the dimensions of the atoms themselves are like the dimensions of the planets and asteroids in comparison with the size of the solar system. All of these computations, whatever their value, are rendered possible by the fact that e is now known.

Now, we know from methods which have nothing to do with the electromagnetic theory of the origin of mass that the excessive minuteness predicted by that theory for both the positive and the negative constituents of atoms is in

fact correct, though we have no evidence as to whether the
foregoing ratio is right.

DIRECT EXPERIMENTAL PROOF OF THE EXCESSIVE MINUTE-NESS OF THE ELECTRONIC CONSTITUENTS OF ATOMS

For more than fifteen years we have had direct experimental proof that the fastest of the α (alpha) -particles, or
helium atoms, which are ejected by radium, shoot in practically straight lines through as much as 7 cm. of air at
atmospheric pressure before being brought to rest.[3] Figs.
1 and 2 show actual photographs of the tracks of such
particles. We know, too, that these α-particles do not penetrate the air after the manner of a bullet, namely, by push-
ing the molecules of air aside, but rather that they actually
shoot through all the molecules of air which they encounter.
The number of such passages through molecules which an
α-particle would have to make in traversing 7 cm. of air
would be about a hundred thousand.

Further, the very rapid β-particles, or negative electrons
which are shot out by radium, have been known for a still
longer time to shoot in straight lines through much greater
distances in air than 7 cm., and even to pass practically
undeflected through appreciable thicknesses of glass or
metal.

Indeed, the tracks of both the α- and the β-particles
through air, can be photographed because they ionize
(knock electrons from) some of the molecules through
which they pass. These ions then have the property of
condensing water vapor about themselves, so that water
droplets are formed which can be photographed by virtue
of the light which they reflect. Thus we see the track of
a very high-speed β-ray. A little to the right of the middle
of the photograph a straight line can be drawn from bottom

---

[3] For details, see *The Electron*, p. 186 and figures.

to top which will pass through a dozen or so pairs of specks. These specks are the water droplets formed about the ions which were produced at these points. Since we know the size of a molecule and the number of molecules per cubic centimeter, we can compute, as in the case of the α-particle, the number of molecules through which a β-particle must pass in going a given distance. The extraordinary situation revealed by this photograph is that this particular particle shot through on an average as many as 10,000 atoms before it came near enough to an electronic constituent of any one of these atoms to detach it from its system and form an ion. *This shows conclusively that the electronic or other constituents of atoms can occupy but an exceedingly small fraction of the space inclosed within the atomic system. Practically the whole of this space must be empty to an electron going with this speed.*

In the case of a negative electron of much slower speed, it will be seen, first, that it ionizes much more frequently, and, secondly, that instead of continuing in a straight line it is deflected at certain points from its original direction. The reason for both of these facts can readily be seen from the following considerations.

If a new planet or other relatively small body were to shoot with stupendous speed through our solar system, the time which it spent within our system might be so small that the force between it and the earth or any other member of the solar system would not have time either to deflect the stranger from its path or to pull the earth out of its orbit. If the speed of the strange body were smaller, however, the effect would be more disastrous both to the constituents of our solar system and to the path of the strange body, for the latter would then have a much better chance of pulling one of the planets out of our solar system, and also a much better chance of being deflected from a straight path itself. The slower a negative electron moves, we can therefore infer, the more is it liable to deflection and the more frequently does it ionize the molecules through which it passes.

But the study of the tracks of the α-particles gives results even more illuminating as to the structure of the atom. For the α-particle, being an atom of helium 8,000 times more massive than a negative electron, could no more be deflected by one of the latter in an atom through which it passes than a cannon ball could be deflected by a pea. Yet toward the end of its path the α-particle does in general suffer several sudden deflections. Such deflections could be produced only by a very powerful center of force within the atom whose mass is at least comparable with the mass of the helium atom.

These sharp deflections, which occasionally amount to as much as 150° to 180°, lend the strongest of support to the view that the atom consists of a heavy positively charged nucleus about which are grouped enough electrons to render the whole atom neutral. But the fact that in these experiments the α-particle goes through 100,000 atoms without approaching near enough to this central nucleus to suffer appreciable deflection more than two or three times constitutes the most convincing evidence that this central nucleus which holds the negative electrons within the atomic system occupies an excessively minute volume, just as we computed from the electromagnetic theory of the origin of mass that the positive electron ought to do. Indeed, knowing as he did by direct measurement the speed of the α-particle, Rutherford, who has greatly helped develop the nucleus-atom theory, first computed, with the aid of the inverse square law, which we know to hold between charged bodies of dimensions which are small compared with their distances apart, how close the α-particle would approach to the nucleus of a given atom like that of gold before it would be turned back upon its course. The result was in the case of gold, one of the heaviest atoms, about 10^{-12} cm., and in the case of hydrogen, the lightest atom, about 10^{-13} cm. These are merely upper limits for the dimensions of the nuclei. The nucleus of the gold atom, exceedingly small though it be, is now definitely known from experiments to be presently described to contain 197 positive electrons and 118 negative electrons.

However uncertain, then, we may feel about the sizes of positive and negative electrons computed from the electromagnetic theory of the origin of the mass, we may regard it as fairly well established by such direct experiments as these that the electronic constituents of atoms are as small, in comparison with the dimensions of the atomic systems, as are the sun and planets in comparison with the dimensions of the solar system. Indeed, when we reflect that we can shoot helium atoms by the billion through a thin-walled highly evacuated glass tube without leaving any holes behind, that is, without impairing in the slightest degree the vacuum or perceptibly weakening the glass, we see from this alone that the atom itself must consist mostly of "hole"; in other words, that an atom, like a solar system, must be an exceedingly loose structure whose impenetrable portions must be extraordinarily minute in comparison with the penetrable portions. The notion that an atom can appropriate to itself all the space within its boundaries to the exclusion of all others is then altogether exploded by these experiments. A particular atom can certainly occupy the same space at the same time as any other atom if it is only endowed with sufficient kinetic energy. Such energies as correspond to the motions of thermal agitation of molecules are not, however, sufficient to enable one atom to penetrate the boundaries of another, hence the seeming impenetrability of atoms in ordinary experiments in mechanics. That there is, however, a portion of the atom which is wholly impenetrable to the alpha particles is definitely proved by experiments of the sort we have been considering; for it occasionally happens that an alpha particle hits this nucleus "head on," and, when it does so, it is turned straight back upon its course. As indicated above, the size of this impenetrable portion, which may be defined as the size of the nucleus, is in no case larger than 1/10,000 the diameter of the atom.

THE NUMBER OF ELECTRONS IN AN ATOM

If it be considered as fairly conclusively established by the experiments just described that an atom consists of a heavy but very minute positively charged nucleus which holds light negative electrons in some sort of a configuration about it, then the number of negative electrons outside the nucleus must be such as to have a total charge equal to the free positive charge of the nucleus, since otherwise the atom could not be electrically neutral.

But the positive charge on the nucleus has been approximately determined as follows: With the aid of the knowledge, already obtained through the determination of e, of the exact number of atoms in a given weight of a given substance, Sir Ernest Rutherford first computed the chance that a single helium atom in being shot with a known speed through a sheet of gold foil containing a known number of atoms per unit of area of the sheet would suffer a deflection through a given angle. This computation can easily be made in terms of the known kinetic energy and charge of the α-particle, the known number of atoms in the gold foil, and the unknown charge on the nucleus of the gold atom. Geiger and Marsden then actually counted in Rutherford's laboratory, by means of the scintillations produced on a zinc-sulphide screen, what fraction of, say, a thousand α-particles, which were shot normally into the gold foil, were deflected through a given angle, and from this observed number and Ruther-

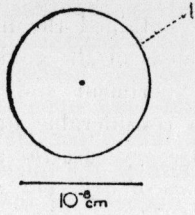

HYDROGEN (1)

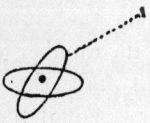

HELIUM (2)

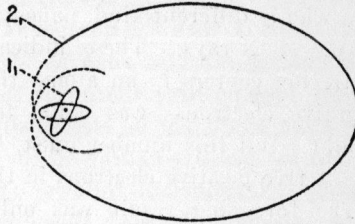

LITHIUM (3)

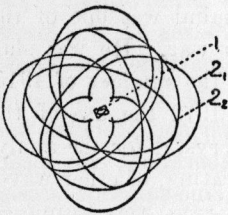

NEON (10)

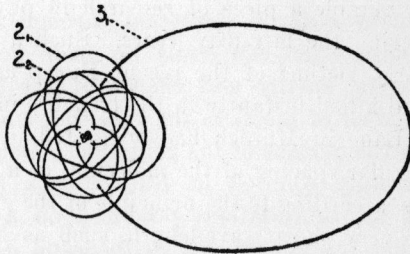

SODIUM (11)

HYPOTHETICAL ORBITS IN WHICH ELECTRONS REVOLVE IN
DIFFERENT ATOMS.

ford's theory they obtained the number of free positive charges on the nucleus of the gold atom.

Repeating the experiment and the computations with foils made from a considerable number of other metals, they found that in every case *the number of free positive charges on the atoms of different substances was approximately equal to half the atomic weight.* This means that the aluminum atom, for example, has a nucleus containing about thirteen free positive charges and that the nucleus of the atom of gold contains in the neighborhood of a hundred. This result was in excellent agreement with the conclusion reached independently by Barkla from experiments of a wholly different kind, namely, experiments on the scattering of X-rays. These indicated that the number of scattering centers in an atom—that is, its number of free negative electrons—was equal to about half the atomic weight. But this number must, of course, equal the number of free positive electrons in the nucleus.

The foregoing result was only approximate. Indeed, there was internal evidence in Geiger and Marsden's work itself that a half was somewhat too high. The answer was in 1913 made very definite and very precise through the extraordinary work of a brilliant young Englishman, Moseley, who, at the age of twenty-seven, had accomplished as notable a piece of research in physics as has appeared during the last fifty years. Such a mind was one of the early victims of the recent European war. He was shot and killed instantly in the trenches in the summer of 1915.

Laue in Munich had suggested in 1912 the use of the regular spacing of the molecules of a crystal for the analysis, according to the principle of the grating, of ether waves of very short wave-length, such as X-rays were supposed to be, and the Braggs had not only perfected an X-ray spectrometer which utilized this principle, but had determined accurately the wave-lengths of the X-rays which are characteristic of certain metals. The accuracy with which this can be done is limited simply by the accuracy in the determination of e, so that the whole new field of exact X-ray spectrometry is made available through our exact knowledge of e. Moseley's discovery, made as a result of an elaborate and difficult study of the wave-lengths of the characteristic X-rays which were excited when cathode rays were made to impinge in succession upon anticathodes embracing most of the known elements, was that these characteristic wave-lengths of the different elements, or, better, their characteristic frequencies, are related in a very simple but a very significant way. *These frequencies were found to constitute a definite arithmetical progression.* It was the square root of the frequencies rather than the frequencies themselves which showed this beautifully simple relationship, but this is an unimportant detail. The significant fact is that, *arranged in the order of increasing frequency of their characteristic X-ray spectra, all the known elements which have been examined constitute a simple arithmetical series each member of which is obtained from its predecessor by adding always the same quantity.*

Photographs show beautifully, first, how the atoms of all the elements produce spectra of just the same type, and, second, how the wave-lengths of corresponding lines decrease, or the frequencies increase, with increasing atomic number.

Since these enormously high X-ray frequencies are due to electrons which are in extraordinarily powerful fields of force, such as might be expected to exist in the inner regions of the atom close to the nucleus, Moseley's discovery strongly suggests that the charge on this nucleus is produced in the case of each atom by adding some particular invariable charge to the nucleus of the atom next below it in Moseley's table. This suggestion gains added weight when it is found that with one or two trifling exceptions, to be considered later, *Moseley's series of increasing X-ray frequencies is exactly the series of increasing atomic weights*. It also receives powerful support from the following recent discovery.

Mendeleéff's periodic table shows that the progression of chemical properties among the elements coincides in general with the progression of atomic weights. Now, it has recently been pointed out that whenever a radioactive substance loses a doubly charged α-particle it moves two places to the left in the periodic table, while whenever it loses a singly charged β-particle it moves one place to the right, thus showing that the chemical character of a substance depends upon the number of free positive charges in its nucleus.

One of the most interesting and striking characteristics of Moseley's table is that all the known elements between sodium (atomic number 11, atomic weight 23) and lead (atomic number 82, atomic weight 207.2) have been fitted into it and there are left but four vacancies within this range. Below sodium there are just ten known elements, and recent study by Mr. Bowen and the author of the spectra of these elements in the extreme ultraviolet has shown that the Moseley type of progression exists here also. It seems highly probable, then, from Moseley's work and that of his successors that we have already found all except four of the complete series of different types of atoms from hydrogen to lead, that is, from 1 to 82, of which the physical world is built, and the discovery of one of these four has recently been announced. From 82 to 92 comes the group of radioactive elements which are continually transmuting themselves into one another, and above 92 (uranium) it is not likely that any elements exist.

Moseley's work is, in brief, evidence from a wholly new quarter that all these elements constitute a family, each member of which is related to every other member in a perfectly definite and simple way. For the succession of steps from 1 to 92, each corresponding to the addition of an extra free positive charge upon the nucleus, suggests at once that the unit positive charge is itself a primordial element, and this conclusion is strengthened by recently discovered atomic-weight relations. Prout thought a hundred years ago that the atomic weights of all elements were exact multiples of the weight of hydrogen, and hence tried to make hydrogen itself the primordial element. But fractional atomic weights, like that of chlorine (35.5) were found, and were responsible for the later abandonment of the theory. Within the past ten years, however, it has been shown that, within the limits of observational error, practically all those elements which had fractional atomic weights are mixtures of substances, so-called isotopes, each of which has an atomic weight that is an exact multiple of the unit of the atomic-weight table, so that Prout's hypothesis is now very much alive again.

So far as experiments have now gone, the positive electron, the charge of which is of the same numerical value as that of the negative, and which is, in fact, the nucleus of the hydrogen atom, always has a mass which is about 2,000 times that of the negative. In other words, the present evidence is excellent that, to within one part in 2,000, the mass of every atom is simply the mass of the *positive electrons* contained within its nucleus. Now, the atomic weight of helium is four, while its atomic number, the free positive charge upon its nucleus, is only two. The helium atom must therefore contain *inside its nucleus* two negative electrons which neutralize two of these positives and serve to hold together the four positives which would otherwise fly apart under their mutual repulsions. Into that tiny nucleus of helium, then, that infinitesimal speck not as big as a pin-point, even when we are in a world which has been swelled ten-billionfold so that the diameter of the helium atom has become a yard, into that still almost invisible nucleus there must be packed four positive and two negative electrons.

By the same method it becomes possible to count the exact number of both positive and negative electrons which are packed into the nucleus of every other atom. In uranium, for example, since its atomic weight is 238, we know that there must be 238 positive electrons in its nucleus. But since its atomic number, or the measured number of free-unit charges upon its nucleus, is but 92, it is obvious that (238 — 92) 146 of the 238 positive electrons in the nucleus must be neutralized by 146 negative electrons, which are also within that nucleus; and so, in general, *the atomic weight minus the atomic number gives at once the number of negative electrons which are contained within the nucleus of any atom.* That these negative electrons are actually there within the nucleus is independently demonstrated by the facts of radioactivity, for in the radioactive process we find negative electrons, so-called beta rays, actually being ejected from the nucleus. They can come from nowhere else, for the chemical properties of the radioactive atom are found to change with every such ejection of a beta ray, and change in chemical character always means change in the free charge contained in the nucleus.

We have thus been able to look with the eyes of the mind, not only inside our atoms which require a ten-billionfold magnification to make them a meter in diameter, but even inside the mere pin-point which is all that the nucleus at the center of that atom becomes with such enormous magnification, and we have been able to count within it just how many positive and how many negative electrons are there imprisoned, numbers reaching 238 and 146 respectively in the case of the uranium atom. And, let it be remembered, the dimensions of these atomic nuclei are about one bil-

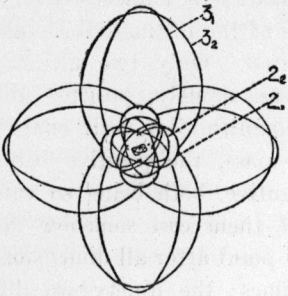

ARGON (18)

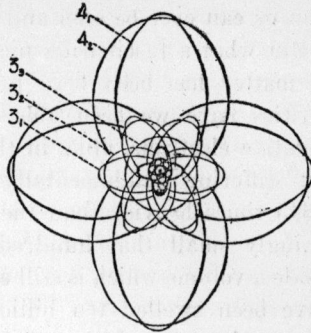

KRYPTON (36)

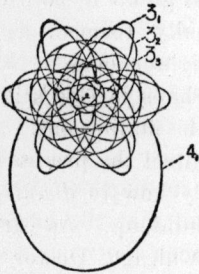

COPPER (29)

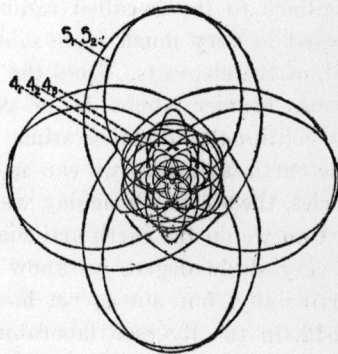

XENON (54)

HYPOTHETICAL ORBITS IN WHICH ELECTRONS REVOLVE IN
CERTAIN ATOMS.

lionth of those of the smallest object which has ever been seen or can ever be seen and measured in a microscope.

But what a fascinating picture of the ultimate structure of matter has been here presented. Only two ultimate entities have we been able to see, namely, positive and negative electrons, alike in the magnitude of their charge, but differing fundamentally in mass, the positive being 1,845 times heavier than the negative, both being so vanishingly small that hundreds of them can somehow get inside a volume which is still a pin-point after all dimensions have been swelled ten billion times: the ninety-two different elements of the world determined simply by the difference between the number of positives and negatives which have been somehow packed into the nucleus; all these elements transmutable, ideally at least, into one another by a simple change in this difference. Has nature a way of making these transmutations in her laboratories? She is doing it under our eyes in the radioactive process— a process which we have very recently found is not at all confined to the so-called radioactive elements, but is possessed in very much more minute degree by many, if not all, of the elements. Does the process go on in both directions, heavier atoms being continually formed, as well as continually disintegrating into lighter ones? Not on the earth, so far as we can see. Perhaps in God's laboratories, the stars. Some day we shall be finding out.

Can we on the earth artificially control the process? To a very slight degree we know already how to *disintegrate* artificially, but not as yet how to build up. As early as 1912, in the Ryerson laboratory at Chicago, Doctor Winchester and I thought we had good evidence that we were knocking hydrogen out of aluminum and other metals by very powerful electrical discharges in vacuo. Without any doubt Rutherford has been doing just this for three years past by bombarding the nuclei of atoms with alpha rays.

How much farther can we go into this artificial transmutation of the elements? This is one of the supremely interesting problems of modern physics upon which we are all assiduously working.

Another fascinating problem! Are the electrons which are held in the outer regions of the ninety-two atoms stationary, or do they revolve in orbits like the planets and asteroids of the solar system about their respective nuclei? We cannot yet answer this question with certainty, but the orbit theory seems at present to be getting the better of the argument. Certainly the wonderful work of Epstein, of the California Institute, in which, by simply applying the theory of perturbations to assumed orbits, he predicted the exact positions and characteristics of all the dozens of spectral lines formed when hydrogen or helium are stimulated to emit light in a strong electric field, is the strongest possible support for the orbit theory. In Figs. 1 and 2 are given the hypothetical orbits in which the electrons, according to Bohr, Nobel prize-winner of 1922, revolve in certain types of atoms. These are the best pictures that we now have of the way in which the electronic inhabitants of this land of Lilliput spend their time.

CHAPTER IV

THE STRUCTURE OF THE COSMOS

By Edwin B. Frost [1]

WE presuppose that it is a cosmos. We examine
the data available to see what we can learn of
its structure. If we are able to show sufficient
evidence that it has a structure which we may regard as
orderly, then we have proven that it is a cosmos. A
cosmos without a structure would be a denial of the definition.

Let us first consider what is implied by structure, or, at
least, how we shall use the term. Our natural human
limitations are so great that it may well appear audacious
for men to attempt to investigate the nature and relations
of the celestial bodies of which we are conscious through
the medium of only one of our senses. Ordinarily, in a
precise investigation of a body, we examine its properties
with the use of all five senses. The sense of sight is the
only one which gives us any consciousness of the existence
of the celestial bodies, excepting, of course, the sun and
moon, and occasional meteorites which reach the earth.
The sun and moon produce tides which may be realized by
the senses of touch and hearing, and the heat of the sun is
recognized very distinctly by the sense of touch.

If we try to characterize the condition of an external
body, we must examine, first, its dimensions, or space relations; second, its mass or quantity of matter; and third,
its time relations. Until recently these have been regarded

---

[1] Director of Yerkes Observatory, University of Chicago.

as distinct and separate. The doctrine of relativity tends to combine these under certain circumstances, but we shall not discuss that theory. The sense of sight, in making us conscious of the existence of a star, also gives us the idea of direction when other objects or points of reference are also visible. It does not give us distance, and the direction inferred is true only if the course of the light has been straight. Our natural experience teaches us that light travels in straight lines, but astronomical observations very promptly show that this is not exactly true except in empty space.[2]

The light which we receive from a star, besides telling us what we regard as its direction (which may be exact or not), also tells us what we call the star's brightness, but gives us no true idea of its distance. It has long been known that "one star differeth from another in glory." If all were actually of the same luminosity, then their apparent brightness would be a measure of their distance, as inferred from the well-established law that the brightness of an object varies inversely with the square of its distance from us. So-called "quantity production" was not, how-

---

[2] Simple measurements prove that at sunset the path of light from the sun's disk is decidedly curved by refraction in the earth's atmosphere, so that when the lower edge appears to touch a marine horizon the upper edge is really just setting, and the lower edge would have set from two to four minutes earlier had the light traveled in straight lines. If something were to happen to the light from the star while coming to us, such as passing through some part of space where the medium is different, we should get a wrong impression of the direction of the star. The fact that light travels with the finite velocity of 186,000 miles per second makes it necessary for us to point our telescopes differently to catch the light of a star than would be the case if light traveled with an infinite velocity, or instantaneously, and if our observations were not made on the surface of a body which is in rapid motion. This is called the aberration of light, and it occurs in several varieties. We cite this to illustrate the fact that our common interpretation of direction is based upon the pure assumption that light always travels in straight lines.

ever, employed in the manufacture of the stars. They are not uniform in size, in mass, or in distance, and they differ greatly in their actual capacity for emitting light, also in the quality of the light which issues from them. It would have been much easier for us to determine many of these important items as to the stars if it had been possible to regard them as manufactured products of perfect uniformity, like modern steel balls for bearings. We shall have to admit that the processes of development of the universe were plainly not designed to simplify the work of astronomers. Some part of our task might, perhaps, have been solved too early, we might say prematurely, and in advance of our correlated knowledge in some kindred branches.

THE MAGNITUDE OF THE COSMOS

These preliminary remarks are intended to confirm our earlier statement that there is an element of the greatest audacity in the attempts of the human mind really to grasp such questions as the structure of the universe. Our units of length had their origin in the size of some man's foot, the length of his arm, or the extent of his stride. As man traveled about, larger units became necessary, such as the distance which a person would usually walk in an hour or a day, units still common in many parts of the world. After the spherical character of the earth was realized, such a unit as the radius of the earth had to be employed, but 4,000 miles was far too small a unit for most astronomical purposes, and we must take as our measuring rod the distance from the earth to the sun, 93,000,000 miles.

In respect to mass, that of an ordinary satellite, like our moon, is 75 million million tons. This might be called an average satellite. The mass of the brain of the average adult human being is about three pounds, but the brain dares to attack problems quite out of proportion to its own mass. In human experience, perhaps, some of the first estimates of mass referred to boats. Now, a modern sea-going monster of steel, a battleship, or a steel office building represents some of the largest products of human construction. The astronomer, however, must take as his unit the mass of our sun, which is 330,000 times that of our earth; and the earth, in turn, has overy eighty times the mass of our moon.

In time, also, we must get away from ordinary human conceptions. One turn of the earth on its axis makes a day; that has always been a convenient unit in human life, and presumably always will be. Another obvious unit is the period of revolution of the earth about the sun, a year. Of course, this is not so significant a unit to a person living on the equator, but to all who live where the seasonal changes are marked, the year is a perfectly natural conception and must remain a significant unit of time for inhabitants of our earth. There is no unit of time longer than the year which is not essentially artificial and arbitrary. The century has no natural significance dependent upon planetary relations. In a sense, it is our largest unit of time for human affairs. But when we examine the history of the earth and of celestial objects, we must at once accustom ourselves to a unit like a million years. In another chapter, the geologist will probably be asking for 3,000 of these units, or three billion years, in order to bring the earth to its present stage of development.

Thus, it is seen that the measures of space, of mass, and of time, required to express properly astronomical conceptions, must be of magnitudes quite transcending ordinary human experience.

THE CONSTITUENCY OF THE COSMOS

The astronomical objects with which we are concerned in the cosmos, in order of closeness of their approach to the earth (the moon excepted), are meteorites, comets, planets, satellites, the sun, the stars, the globular star clusters, the nebulæ.

Meteorites

These are bodies of a stony or metallic nature, composed of 27 chemical elements which are all familiar to us on the earth's surface. They bring us from outside the earth no new element. The well-known swarms of meteorites have orbits similar to those of comets and they appear to represent a stage in the disintegration of a comet. Recent investigations strongly indicate that comets belong to our solar system and do not come to us from beyond it; hence we draw the inference that meteorites are samples of our solar system and do not represent the material of stellar systems, although, as we shall show later, they might equally well be samples from external systems, so far as their chemical constitution is concerned.

Comets

These are unsubstantial objects, often of immense size, which may approach nearer the earth than any of the planets or asteroids of our system. We know that when they are in the range of observation from the earth, they are moving under the sun's attraction in orbits that are elliptical, with the sun at one focus, although for convenience in computation their paths are often treated as parabolic. A parabola is an ellipse whose long axis is infinite. That the orbit should actually be parabolic would be the purest accident. The planets also, particularly Jupiter and Saturn, exert a strong pull on the comets, and they have profoundly modified the primitive orbits of many of them. They might even convert the orbit into an hyperbola, in which event the comet would never return.

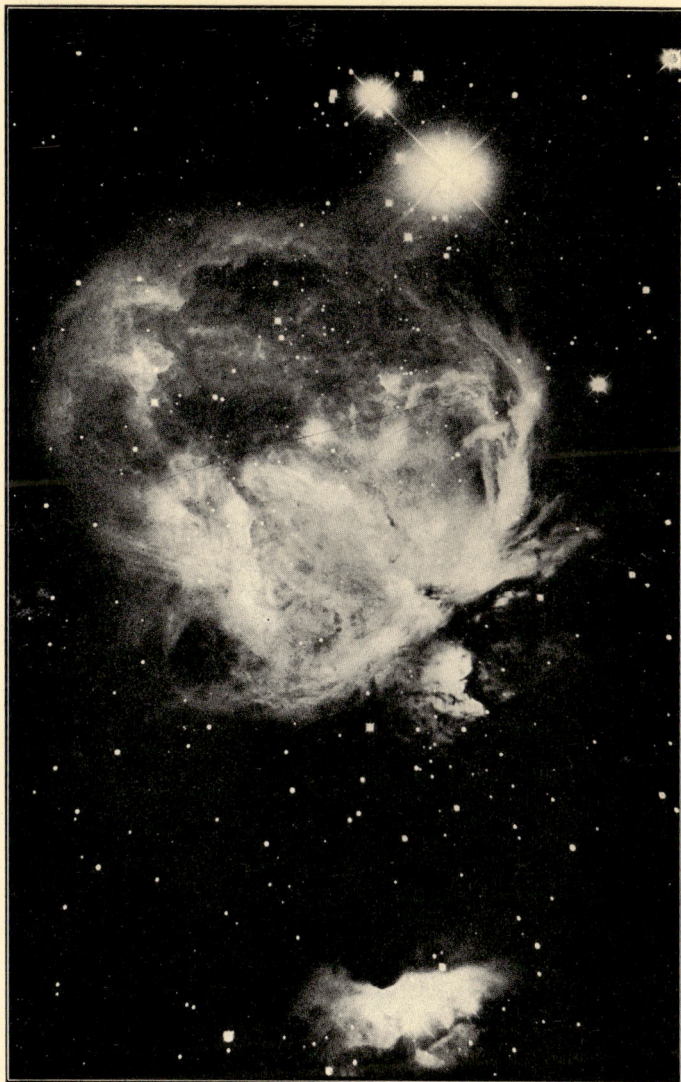

THE GREAT GASEOUS NEBULA IN ORION.

This was photographed by G. W. Ritchey with the 24-inch reflector of the Yerkes Observatory. This enormous object, writhing in the commotion of its gaseous constituents, may well represent chaos. Its distance is probably of the order of six hundred light years, and its greatest diameter is some six or eight light years.

Thus, if the comet describes its orbit about the sun in a period of from five to seven years, we infer that it is brought into this position by the dominant action of Jupiter. If the period is about fourteen years, we regard Saturn as the planet which coerced the comet into a more or less permanent and smaller orbit than it once had. Neptune is held responsible for two or three comets, having a period of about eighty years, the most celebrated of which is Halley's.

Comets disintegrate and disappear before our eyes, fragments sometimes separating from the main head for reasons not fully known, but often assigned to an electric repulsion. The main tail of a comet always points away from the sun, and the cause of this has been traced to the "pressure of light" or reaction produced upon very fine particles which absorb the energy of the sunlight. This pressure of light has been demonstrated and measured in the laboratory. As a result of this force and other forces less understood, comets' tails disport themselves in a manner almost unseemly for well-ordered astronomical objects: the tail may twist, or break off, or change its direction in a very short time, even in an hour.

The spectrum of a comet indicates no unfamiliar chemical elements, being predominantly carbon, nitrogen, and hydrogen, and these in molecular combinations known as hydrocarbons. In fact, the spectrum of the gaseous portion of an acetylene flame matches that of a comet quite closely. Cyanogen gas, which is, of course, a deadly poison, is also present in the tails of some comets, and this fact has given rise to some popular, but superfluous, fears that when a comet's tail sweeps too near the earth the population might be suddenly asphyxiated, even as were the inhabitants of Saint Pierre on the Island of Martinique when the deadly gases rolled down the slope of Mont Pelée. The gases forming the tail of a comet are of extreme tenuity—

perhaps almost equal to the vacuum of an incandescent lamp, and it is very doubtful whether we should know it if a comet's tail actually should envelop the earth. There is a reasonable degree of probability that the earth was lightly brushed by the tail of Halley's comet on May 18-19, 1910. We cannot say with certainty whether or not this occurred, but no really noticeable effects could be detected in our atmosphere. Meteorites and comets are the rubbish of the solar system, but aside from their being interesting and even sometimes spectacular objects, they serve a useful purpose in testing out for us the condition of things in inaccessible parts of the solar system. They can go on their erratic ways where no planet can follow, even daring to graze the sun, and yet come out unscathed. They can test for us the mass of other planets, for if they approach close enough, the comet's orbit is notably changed, giving us a measure of the gravitational pull, and hence the mass, of the planet.

Planets

From the human point of view, a planet, and in particular a geoid, is the most indispensable of the celestial bodies. We can use all of our senses in the study of the earth, but for the other planets we are again limited to the sense of sight. Our sun is encircled by a group of at least eight of them, besides a thousand or more fragmentary planets, or asteroids, occupying a region between Mars and Jupiter. It is an interesting question whether planets are the normal attendants of a star. It is hardly conceivable that our sun should be the only star so provided with a retinue of attendants. But at present, we cannot give objective evidence of the existence of any planet attached by gravitation to any other star than our sun. It is true that among the so-called "spectroscopic binaries," or double stars so close that they cannot be separated by any telescopic power, but only with the spectroscope, there are found to be many pairs of twin stars. One of these may be much fainter than the other, or perhaps even non-luminous and eclipsing the brighter one at every revolution, or ordinarily every few days. The faint companion in these cases, however, is too nearly of the magnitude of the primary to be called a planet, for in our solar system the largest of the planets, Jupiter, has a mass less than 1/1,000 that of the sun. If the nearest star had a planet the size of Jupiter, and at its distance from the sun, we could not discern it with any of the most powerful instruments at our command at the present day. The reader is, therefore, quite as competent as the astronomer to form an opinion as to the probability that our star, the sun, is the only one to have a family of planets.

The question of the habitability of our planets, although far too much exploited by certain misrepresentatives of the press, is one of legitimate interest to thoughtful persons. Unfortunately, no positive answer can be given in respect to the only two which could be considered, namely, Mars and Venus. It would be rash to say that those two planets could not support organic life—in fact, it seems reasonably possible that Mars should do so, but its water supply is evidently so limited that we could not expect a high degree of development of such life. Venus seems to be surrounded by a thick envelope of atmosphere or clouds which prevents us from obtaining any view of the planet's surface. If it is less likely that organic life now exists on Venus, we cannot assert that this planet may not at some future time become suitable for such life, if it has not been so at some time in the past. We venture a speculation that organic life begins on a planet just as soon as the conditions are suitable, and as simply and quietly as the hepatica comes to bloom when conditions are right on our northern hillsides.

Satellites

The planets of our solar system, for the most part, are themselves centers around which smaller bodies circle at precisely the speed required by the law of gravitation. So far as is yet known, Mercury and Venus are exceptions, being unattended. The satellites of our planets, with a few exceptions, move around the planet in the same direction, west to east, that the planet turns on its axis, and that the planet moves in its own orbit around the sun. The planes of the satellites' motions, also, fall near to those of their orbits and to that of the earth's orbit, called the ecliptic. Saturn goes beyond the other members of our system in the matter of these satellites, for it has an uncounted number of them forming the rings which have made the planet an object of especial beauty and interest ever since they were discovered by Galileo in the first uses of his telescope. This sheet of minute satellites is probably of the order of twenty-five miles in thickness, but has an extreme diameter of 173,000 miles. The rings were a puzzle to astronomers, mathematicians, and physicists, until, in 1895, it was proven by the spectroscope in the hands of the late J. E. Keeler that at every distance from the center of Saturn, the ring was moving at exactly the velocity demanded by the harmonic law of Kepler, just like the other satellites of the Saturnian system.[3] Saturn's rings, as well as the satellites, were items which played a considerable part in the

---

[3] The inner portion of the ring rotates far more rapidly than the outer portion, in such a fashion that the fraction $\frac{a^3}{t^2}$ is the same for each satellite, and for every part of the rings. Here a represents the distance from the center of Saturn, and t the period required for one revolution of the particle. This is the harmonic law, and a consequence of the Newtonian law of gravitation. The formula applies to the planets in their circuit around the sun, and in that case a represents the mean distance from the planet to the sun, and t the planet's period of revolution about the sun.

theories of Kant (1755), and Laplace (1796), as to the origin of the solar system, and long known as the Nebular Hypothesis. It was, however, impossible for them to know whether the rings were a solid or a fluid sheet, or that they were in actual motion.

The Sun

Much has been written about the dependence of the inhabitants of the earth upon the sun for light and heat, and hence food and continued life; for mechanical power, whether from the steam engine, from the water-fall, or from the wind; and, in fact, for all of our human activities; nevertheless, this dependence cannot be impressed upon us forcibly enough. We should remember, also, that all of the æsthetic beauty of color on the petals of a flower or the wing of a bird is but the reflected glory of the solar rays. Verily, the dependence of us terrestrials upon the sun cannot be exaggerated.

The sun, our star, is a gaseous sphere about 864,000 miles in diameter, not appreciably flattened at the poles. It turns on an axis inclined about 7° to the ecliptic, in a period of 25.4 days, as determined from observations of the passage of spots across its surface. It contains about 330,000 times as much matter as the earth, and its volume, or bulk, is about 1,300,000 times that of the earth. Consequently, its average density, which is its mass divided by its volume, both taken with respect to the earth, is about one-fourth that of the earth's average density. This is about 1.4 times the density of water.

The apparent surface of the sun, known as the photosphere, is covered with a very fine mottling, visible in the telescope in spite of its brilliance; and it radiates as if these mottlings were great clouds of liquid matter. However, it is impossible to believe that matter could exist in the liquid form at the high temperature evidently prevailing in the photosphere; in fact, it is hard to imagine how a gaseous ball of high temperature can present a disk such as we find. The luminosity of the sun is enormous, according to all terrestrial standards, and a statement of its candle power has little meaning for us. The darkest part of a sun-spot is even brighter than an arc lamp, appearing dark merely by contrast.

The late Professor Samuel P. Langley, of the Smithsonian Observatory at Washington, devised an exceedingly sensitive method of measuring accurately the amount of radiation which the earth receives, per square meter of its surface, from the sun. The process has been improved and simplified by his successor, Charles G. Abbot, so that the "solar constant" is determined daily at two rainless stations, in Chile and Arizona.[4] One of the most interesting results of this work is the demonstration that the "solar constant" is variable by several per cent, in exceptional cases as much as seven per cent. The significance of this for the earth is very great, but it cannot yet be interpreted with any sufficient accuracy in forecasts of terrestrial weather.

The sun is not immaculate. Spots appear upon its surface, occasionally of a size large enough to be visible to the naked eye. There are many records of them in the Chinese annals running back to the early centuries of our era. Their periodicity was discovered from systematic

---

[4] The average value of the solar constant as derived from 1244 measurements at the stations of the Smithsonian Observatory for the period 1912-1920 is 1.946 calories per square centimeter of the earth's surface, per minute, when the sun's rays are perpendicular, and due allowance has been made for the absorption of the heat in our atmosphere. The unit of heat known as the calory is the amount of heat required to raise by one degree Centigrade one cubic centimeter of water, which is one gram. A better idea of the amount of this heat may perhaps be gained from the fact that it would be sufficient to melt a sheet of ice eight and one-half inches thick over the whole earth every twelve hours.

telescopic observation about eighty years ago. The length of the cycle is, on the average, 11⅛ years, but may be as short as seven, or as long as seventeen, years. In 1918 we went through a maximum of solar activity; spots were larger and more numerous than for some years previous, and the eruptions seen with the spectroscope at the sun's edge were similarly more frequent and more violent. In 1923, few spots were visible, and most of them small, and the solar prominences were relatively infrequent and of small size. No cause can be at present definitely assigned for this intermittent activity of the sun; although often asserted, it has not been demonstrated that the gravitational action of the planets is responsible. We cannot doubt that these solar variations have an effect upon the circulation of the earth's atmosphere, and hence upon our weather, but the relations have not yet been definitely established.

Sun-spots are considered to be whorls in the circulation of the solar atmosphere, having some analogy to the ordinary storms in the earth's atmosphere, as indicated on the weather maps. They occur only within certain zones of latitude on the sun's surface, being seldom seen more than 35° from its equator. Accurate observations of their position have shown that the rotation of the sun is fastest at its equator (27.25 days as seen from the earth), while it is about 1½ days slower at solar latitude 30°. This shows that the sun does not rotate as a solid body.

Our sun is spectroscopically classified as a dwarf of type G0, which means that it is a comparatively small yellow star having many dark lines in its spectrum. These dark lines are really gaps or intermissions in the continuity of the ribbon of color extending from the scarlet to the violet. A very precise comparison of the positions of the dark lines in the solar spectrum with the bright lines emitted under laboratory conditions by specimens of the elements, furnishes us with an exact identification of the origin of the lines in the solar spectrum. It is thus found that the sun contains forty-eight of the chemical elements, or something more than half of those found on the earth. The spectral lines do not suggest the presence in the sun of any element not occurring in the earth, except coronium, which exhibits bright lines in the outer envelope of the sun, or corona, observable only at times of total solar eclipse. Helium, however, was discovered spectroscopically in the sun in 1868, but was not found in the earth until 1895. The elements certainly known to be present on the sun are among those with which we are most familiar on the earth, such as the gaseous elements hydrogen, helium, oxygen, and nitrogen, and the metals iron, nickel, titanium, manganese, calcium, sodium, magnesium, etc., together with carbon and other nonmetallic elements.

While the presence in the spectrum of a celestial body of the lines characteristic in the laboratory of a given element is a definite proof of the presence of that element in the body, the converse is not true. The absence of the lines of an element is no proof that the element itself may not be present, but under conditions of excitation or ionization insufficient to bring out the lines.

The precise knowledge of the sun's radiation is of much importance in our philosophy, as to its cause and continuance. Twenty years ago it was commonly thought, as a result of the studies of von Hemholz, Lord Kelvin, and others, that the sun's contraction produced sufficient energy to balance exactly the losses due to its continuous radiation. But looking backward and forward, according to this theory a beginning and an end could be seen in a relatively short period of time. A gaseous ball of the size of our sun, contracting under its own gravitation, would not maintain its supply of heat indefinitely; far from it. In the course of a few million years, possibly from twenty to fifty, the vitality of our orb of day would be exhausted. Its career as a living star could hardly be longer than the last named figure. That was perhaps long enough for the modest ideas of time which we had at the beginning of the twentieth century. Now, stellar evidence indicates immensely longer periods for the life of a star, and, as we have already said, geologists demand for the earth alone some thousands of millions of years. The primary source of energy of a system must certainly long precede the origin of an inconsequential subsidiary like our little earth, and must survive long after our planet has run what we are pleased to call its course, that is, has ceased to be suitable for the support of human life as we know it. Otherwise, our view would be so anthropomorphic as to make the sun dependent upon the earth, rather than the reverse.

It is natural to speak of the temperature of the sun and of the stars, but any statement on this point must be qualified rather precisely. Considerable evidence has recently been brought forward to show that the temperature of the interior of the sun or of a star may be very high—perhaps reaching millions of degrees for the hottest stars—and it is natural that it should be different in its various strata. Accordingly, we use the term "effective temperature," meaning thereby the temperature that a perfectly radiating surface, of the same size as that of the sun, would require in order to give out the radiation which the sun emits. Tolerably accordant results have been obtained by a number of investigators, using quite different means of measurement, and we may take the value as 6,000° C or 11,000° F. as representing the best measurements. We shall see that this value is intermediate between that found for the hotter and the cooler stars, agreeing well with the results found for other yellow stars of the spectral type of our sun.

The comparatively recent discovery of radioactivity and some of its possible consequences have revealed to us new presumptive sources of solar energy. Our increasing realization of the enormous stores of energy within the atom have also given us larger possibilities for the perpetuation of the sun's radiation. Perhaps the interior of a star is a power-house for the conversion of atomic energy into the energy of heat. Thus, the new discoveries free us from limitations imposed by merely gravitational sources of energy and let us think in terms of time consistent with that which the stellar universe implies.

We shall shortly see that our appreciation of spatial relations does involve relations of time because we get no instantaneous picture of our cosmos. An explosion on the sun is eight minutes past when we see it; the blazing up of a nova, or new star, may often be belated by a millennium, on account of the comparative slowness of the passage of light (11 million miles per minute)! The appearance of such a nova in a spiral nebula, where novæ most frequently occur, may not be recorded on our photographic plates for fifty thousand years after the event. A casual cataclysm in a spiral nebula might even require a million years before the speeding waves of light could transmit the news to our post of observation. Thus we are forever denied a contemporary cross-section of the universe. An immediate inference is that the stages of cosmic evolution are very long and that the active or self-luminous phases of the life of celestial objects must be far greater than we thought a few years ago. But we shall also see before we are through with this chapter that some processes occur with catastrophic suddenness.

We have discussed the sun at some length because we may safely regard it as a fairly typical star. A comparative nearness of 93,000,000 miles permits us to examine it to an extent which is entirely impossible for any other star, but we may safely apply to the stars of similar spectral type the information which we have gained from the sun.

The Stars

From the terrestrial point of view, there is a vast gap between the distance of the sun and that of the stars. Our nearest known stellar neighbor, Alpha Centauri, is 275,000 times as distant from us as the sun. This distance may best be expressed as 4.3 light years. Leaving the three stars which compose the system of Alpha Centauri, the next nearest star, so far as is at present known, is Barnard's "run-away" star, at a distance of 6.0 light years. This star has a motion in space of 87 miles per second. The spherical space having our sun at the center, and a diameter of 20 light years, contains only 10 other stars; in a sphere within a diameter of 65 light years 104 stars are known to exist. Our first imaginary sphere would, therefore, including our sun, have one star in every 388 cubic light years of space, while in the larger volume there would be nearly 1,400 cubic light years of space as the average elbow room for each of the 105 stars.[5]

The Distances of the Stars.—Direct trigonometric methods of determining the distances of stars fail entirely for distances greater than about one hundred light years, and indirect methods have to be employed. It has recently been found possible to obtain a measure of the absolute magnitude of stars from a comparison of the relative intensities of certain pairs of lines in their spectra. Absolute magnitude here means the brightness that the star would have if it were placed at a standard distance of 32.6 light years from the observer. It is obvious that the brightness of a star will decrease as its distance from us is increased;

---

[5] If these dimensions seem small, let the reader bear in mind that one light year is nearly 6 million million miles, or the distance light travels in a year at its usual rate of 11 million miles per minute, and this number must be cubed to give the number of cubic miles in a cubic light year—in round numbers, $200 \times 10.^{27}$

in fact, the brightness varies inversely as the square of the distance. As a class, the fainter stars will evidently be more distant than the bright stars, however much the range of individual brightness may be. If the spectral characteristics of the stars, which depend upon their physical conditions, give us a measure of the brightness which the stars would have if situated at the standard distance, then it is evident that if we could measure their actual brightness with a photometer we may determine conversely what their distances must be to produce the brightness actually observed.

The Brightness of the Stars.—The term magnitude is a somewhat unfortunate one, as it does not refer at all to dimensions, but merely to brightness. Following the plan of Hipparchus, it came to be a usage to divide the stars visible to the naked eye into six grades of brightness, or magnitudes, the brightest being called of the first magnitude and the faintest of the sixth. When the light of these stars was later measured accurately with photometers, it was found that the average star of the sixth magnitude was only one hundredth as bright as the average star of the first magnitude; hence it followed that the ratio of brightness from one magnitude to another must be the fifth root of 100, or 2.5+. The average fifth magnitude star is therefore 2.5 times as bright as the average sixth magnitude star; one of the fourth magnitude is 2.5×2.5 brighter than one of the sixth magnitude. A star of the eleventh magnitude would be 2.5^{10} or 10,000 times fainter than the average star of the first magnitude. The faintest star that can be reached with our largest modern reflecting telescope has probably about the twentieth or twenty-first magnitude. This difference of twenty magnitudes would therefore correspond to a ratio of 100 million, which would be the number of times that the brightness of the average first magnitude star exceeds that of a star of magnitude 21.

The Number of the Stars.—A keen eye, without tele-scopic aid, may discern on a perfectly clear, but moonless night, about 2,500 stars at one time. The number of stars over the whole sky, which could be seen by a good eye in the course of a year, is a little less than 5,000. The query naturally arises whether the number of stars is infinite. The answer from most astronomers who have studied this question would probably be "No." If a count is made of the number of stars of each magnitude, it will be evident that the number increases from three to five-fold as we pass from one magnitude to the next fainter, but this ratio does not continue after we reach the eleventh magni-tude, and the increase is scarcely more than two-fold as we pass from the fifteenth to the sixteenth magnitude. Thus, from Kapteyn's counts we find that the number of stars of the sixth magnitude is 4,720, while the number of the group averaging the seventh magnitude is 15,000, or 3.2 times as many. Of the tenth magnitude there are enumerated 379,000, and of the eleventh, 1,020,000, which numbers are in the ratio of 2.7; of the fifteenth magnitude there were estimated to be 27,500,000, and of the sixteenth, 57,000,000, giving a ratio of 2.1. This diminution in the increase with magnitude, therefore, points to a finite limit to the number of stars. We may further cite the old argu-ment that if the number of stars were infinite, the whole heavens should be ablaze with light at night. Careful investigation has been made of the question whether there is an absorption of the light of the stars in space. If such exists, the most distant stars would suffer the greatest loss in light, and the above conclusions would be contravened. No progressive absorption in space can be detected by any modern methods of investigation, although we now recog-nize the obstruction by dark absorbing matter in certain definite regions of the sky.

Estimates of the total number of stars involve extrapolation and assumptions which cannot yet be proved, so that we can say only that the number of luminous stars is in excess of hundreds of millions, but probably not of the order of many thousands of millions. It is an interesting fact that the illumination that we get from starlight on a clear, moonless night is to the extent of 95 per cent derived from stars which we cannot see with the naked eye, in other words, from stars fainter than the sixth magnitude.

Distribution of the Stars.—When the number of stars in a given area (as a square degree) is counted over the whole heavens, it is apparent that there is a great increase in the number along the zone which we recognize as the Milky Way. Thus, Sir William Herschel, who was the first to make these counts, found almost thirty times as many stars per unit area along the central line of the Milky Way as he did at a point 90 degrees distant, at the poles of the Milky Way. It is evident that the stars of our stellar system are largely concentrated in the Milky Way, and that our stellar system has the shape of a disk with its diameter some ten times its thickness. From some studies of the Milky Way it has been inferred that it may actually be spiral in form, but we are, of course, poorly placed to have this structure brought out. The general form of the Milky Way is that of a great circle cutting the plane of the ecliptic at two opposite points, and at an angle of about 60 degrees with it. Its north pole is situated in the constellation of Coma Berenices. The width of the belt of the Milky Way varies from a few degrees in its narrowest place to as many as 30 degrees where it is widest. It consists of two branches in the region from Cygnus to Scorpio, a portion best visible in the skies of our summer evenings. It varies greatly in brightness at different points and shows notable dark places to the naked eye. Its flocculent character is brought out on the wonderful photographs made by the late Professor E. E. Barnard of the Yerkes Observatory, who was a pioneer in the application of photography to the study of the Milky Way. The so-called star cloud in Scutum (Plate xxx) gives an impression of the immense wealth of stars concentrated in its brightest parts.

For many years it was thought that our sun was rather near the center of our galactic system, but recent studies displace it considerably from that position. The extent of the Milky Way is more in doubt at present than it was thirty years ago when the view was general that a diameter of some 30,000 light years was regarded as large enough. Evidence has been adduced by Shapley to show that this should perhaps be increased from fivefold to tenfold. We shall refer to this again in the discussion of the spiral nebulæ.

Quality of the Light of the Stars.—The light of the stars can give us information only in a limited number of ways: first, it tells us the direction of the stars, and hence, if the direction changes, we may infer their motion; secondly, measurements may be made of the brightness of the stars as this affects the eye or the photographic plate. We may also examine the quality of the light in respect to the different colors or rays present. The absence of certain radiations gives us the clew to the chemical elements present in the stars. We shall see that the exact measurement of the intensities in the different colors will lead to a determination of the effective temperatures of the stars. The quality of the starlight is investigated by the spectroscope.

Variable Stars.—Careful study of the brightness of the stars, both by visual estimates and by measurements with photometers, has shown that very many of the stars fluctuate in their light, some five thousand being already recognized as certainly variable. In some cases the star's light will be found to wane at regular intervals and after a few hours or days to return to its normal brightness. A well-known example is the star Algol, which is proved to be eclipsed by a dark companion every two days and twenty-one hours, losing five-sixths of its light at the time of minimum. Other stars, ordinarily faint, rise to a maximum at more or less regular intervals, from some unknown cause, but evidently any eclipse is excluded as an explanation of their behavior. There are many red stars which have a period of about eleven months, in which they may change their brightness by eight or ten magnitudes, some of them being ten thousand times as bright at maximum as at minimum. A classical example is Omicron Ceti, also known as Mira, the Wonderful, which for over three centuries has been known to vary from the ninth or tenth magnitude at minimum to a maximum sometimes as bright as second, and sometimes attaining only fourth magnitude, with an average period of 331 days. There are many other types of variation, evidently due to very different causes. The periods range from as short as three hours and fifteen minutes to as long as nearly thirty years, while for some of the irregular variables the periods may reach centuries.

The stars already discovered to be variable represent only a small fraction of the whole number existing. We have seen that our own sun varies in the amount of its radiation, but by an amount which would escape detection from another star. Accordingly, inconstancy seems to be a frequent characteristic of the stars, despite the conventional reference by poets and others to the steadfastness of their shining.

Size of the Stars.—We speak of the sun as a rather typical yellow star, but of the "dwarf" class. No star is large enough to present a disk visually measurable, as such, in our largest telescopes. From the distance of Alpha Centauri the diameter of our sun would subtend an angle of only 1/275,000 of its present angular diameter, or seven thousandths of a second. This would be more than ten times less than the resolving power of our largest telescopes. For a star very much larger than our sun, however, and comparatively near, the diameter approached the possibility of perception by the method of interference of light waves proposed by Professor A. A. Michelson, years ago. By the application of this principle, with further improvements devised by Professor Michelson, it became possible in 1920 to determine the diameter of Betelgeuse, or Alpha Orionis, with an 18-foot interferometer attached to the 100-inch reflector of the Mount Wilson Observatory. The surprising result was obtained that this star's diameter is 300 times that of our sun, the star's distance being taken, from the best recent measures, as 210 light years. Similar measures with the interferometer, of the star Antares (Alpha Scorpii) gave a diameter even greater. These are giant stars, and their volume or bulk, from the data at present available, would be respectively 27 million and 40 million times that of our sun. We therefore conclude that the range of size of the stars is very great, from a volume of somewhat less than that of our sun to the enormous dimensions above given.

Mass of a Star.—The mass of a star, or the quantity of matter which it contains, however, appears to be confined within a much smaller range, perhaps from one-tenth that of our sun to ten or twenty times that of our sun, and exceptionally fifty times. The mass is found, as compared with that of our sun, when we can determine the orbit of a binary star and get the relation of its separation and period in comparison with the same relation for the earth in its orbit around the sun, the law of gravitation being rightly assumed as applying to the star. The density of a star, which is its mass divided by its volume, will evidently also have a very great range, being excessively small in the case of super-giants like Betelgeuse and Antares; in fact, being comparable with the vacuum which is maintained in the bulb of an electric lamp. The densest star thus far known, a dwarf in the constellation Coma, is from two to three times as dense as the sun, or about seven-tenths as dense as water.

Stellar Spectra.—The words "spectrum" and "spectroscope" may have an ominous sound to some readers, but we shall try to speak simply of them. If we are to study the different colors in the light of a star, we must have them spread out before us, and not concentrated and overlying each other in a point or dot. This separation can be accomplished most simply by allowing the light of the star to pass through a prism. The violet rays will be bent or refracted most, and the red rays least, by such passage. The blue, green, yellow, and orange rays will be bent to an intermediate degree in that order. Thus, instead of having a dot or a point of white light, we get a thin ribbon of colors. If we place a prism over the object glass of a photographic telescope, and make a suitable exposure, we find that every dot shown on a photograph taken without the prism has now become a slender band on the photograph taken through the prism. This is the simplest form of spectroscope, and has proven very efficient for dealing with large numbers of stars, particularly in the work of Miss Annie Cannon at the Harvard Observatory. For convenience in examining stellar spectra of this sort, it is customary to widen the band by allowing the telescope to drift slightly by altering the rate of its driving clock. It will be found on these widened spectra that there are certain occasional gaps or intermissions in the band, perpendicular to the length of the spectrum. These are the dark lines of the spectrum, and from their study we may learn the chemical elements present in the atmospheres of the stars.

When the spectrum of a white star is photographed in this way, a few strong lines are seen, and it can be readily proved in the laboratory that they are due to the absorption by a stratum of hydrogen in the outer atmosphere of the star. The majority of the brighter stars are white and exhibit this type of spectrum. The yellow stars, like our sun, show a large number of narrow lines, and two very strong dark lines near the violet limit of the visual spectrum. Analysis with powerful spectroscopes discloses that these fine lines are chiefly due to the metallic elements, as was the case in the solar spectrum, and the two very strong lines known as H and K at the violet end are due to absorption by calcium vapor.[6]

The stars having an orange or reddish color give, in addition to the fine lines, a striking series of dark bands which produce the effect of a columnar or fluted structure. These bands have been identified with the absorption due to a gaseous compound known as titanium oxide, titanium being a metallic element very abundant in the earth's crust and in the stars generally. Stars of a deep red color, which are likely to be variable in their brightness, also exhibit a band spectrum due to a combination of carbon and hydrogen known as the hydrocarbon spectrum. The color of a star gives, of course, an indication of its temperature, for from our experience with hot metals we at once infer that a red-hot substance is cooler than one that is white-hot. Further, compounds like those mentioned can exist only at comparatively low temperatures, thus fully confirming the evidence along other lines to the effect that the deep-red stars have the lowest effective temperatures.

For the minute study of stellar spectra, spectroscopes much more powerful than the simple form described above are used, the light of the star falling upon a narrow slit

---

[6] Absence from the spectrum of a celestial body of the lines of any element does not prove that the element is not present. It merely indicates that the conditions are not such that the electrons can excite the vibrations responsible for the lines. Conversely, however, the case is different: the presence of the lines characteristic of a given element is a sure proof of the presence of that element.

at the focus of the telescope, and then passing through a train of prisms, and a camera lens which throws the image of the spectrum on the photographic plate. In this arrangement of apparatus it is customary to project also upon the slit the light from an electric spark leaping between metallic terminals, or from a vacuum tube through which a current is passed. Thus there is photographed above and below the star spectrum that of some well-known chemical element used as a standard of reference. The lengths of the rays of the artificial spectrum used for comparison are generally known within an accuracy of one billionth of a millimeter, or one twenty-fifth of a billionth of an inch. The length of a wave of green light is about 1/50,000 of an inch.

When the spectra of the stars were examined systematically fifty years ago, they were divided into four groups, known as Secchi's types, as described above. More minute analysis, with more powerful apparatus, has shown the existence of important sub-classes, and it has been the custom in recent years to follow the Harvard classification, developed by the late Professor E. C. Pickering, in which arbitrary letters represent the different classes. The sequence is unmistakably shown in any extended collection of spectra taken with proper instruments, and more than fifty subdivisions can be distinguished.[7]

---

[7] According to this system, Class O represents very white stars which may have bright lines as well as dark; Class B comes next, being characterized by the presence of a number of lines of helium in addition to those of hydrogen; Class A then follows substantially as described for the first of Secchi's types, but the fine metallic lines become apparent with an apparatus of higher power. The next letter employed is F, which applies to the stars showing a trace of yellow, such as Procyon. The solar spectrum comes next and is denoted by the letter G. K is used for stars having more pronounced absorption, with many fine lines, and includes such stars as Arcturus and Aldebaran, in which there is a distinctly orange tinge. M is the next letter employed, corresponding to Secchi's

The spectrum of a purely gaseous body like a gaseous nebula differs from a star in that it exhibits only bright lines on a dark background instead of dark lines on a background of the chromatic spectrum. Were a more brilliant incandescent body to be placed behind the gas giving the bright lines, they would then change to dark. Just as the bright lines at the edge of the sun are turned to dark when projected against the brilliant photospheric disk, there are many stars which show in their spectrum both bright and dark lines, in accordance with peculiarities in the character and extent of their atmospheres.

Careful measurements of the intensity of the radiation in different parts of the spectrum of a star, just as in the case of the sun, have led to a knowledge of their effective temperatures. It is thus found that the temperature of the very hottest stars, which we call blue in color, and which are of Class B, is about 17,000° Fahrenheit, while in the white stars of Class A, the temperature is somewhat less, 15,000°. Classes F and G have an average temperature of about 10,500,° while for the cooler M stars the temperature is found to be 6,000°.[8] It is not at present possible to assign a temperature to the gaseous nebulæ. From certain considerations this would appear to be high, but we are unable to understand how, if high, it can be maintained.

---

third type, while N corresponds to the fourth. R and S are extensions of types M and N, which in some respects appear collateral rather than successive. Intermediate types are designated by numbers which are supposed to represent, on a scale of ten, the position of a star between one letter and the following. Thus, A5 denotes that a star is half-way between A0 and F0; G5, that it is half-way between G0 and K0.

[8] These temperatures are reckoned from the absolute zero, which is 459° below the Fahrenheit zero. This number of degrees would, therefore, be substracted from the above figures, as was the case with the sun, earlier in the chapter, if it is desired to reckon the temperatures from the ordinary Fahrenheit zero.

Speed of the Stars in the Line of Sight.—The spectroscope is very remarkable for its revelations, but in no respect more so than when it furnishes us the speed with which the stars are moving toward or from us in the line of sight. The principle involved, known as that of Doppler, who was the first to deduce it, is simple. If an object emitting monochromatic waves rapidly approaches the observer, then more waves per second will be received by him, and he will get the effect of a slight shift of the waves toward the violet end of the spectrum. If the object is receding, he will get less waves per second than if the object were at rest. The effect would be to shift these waves toward the red end of the spectrum. The same effect is produced if the observer himself is moving rapidly with respect to the luminous object. The artificial comparison spectrum impressed on each plate, adjacent to the stellar spectrum, will give the position which the rays should have when emitted from a body at rest. The measurement of this slight displacement, seldom as much as a thousandth of an inch, thus gives the velocity of the observer with respect to the star or nebula, but it obviously measures only that component of the star's speed which is in the line of sight. If a body is approaching us with a speed of one thousandth that of light, we get the effect of one thousandth more vibrations per second, and the wavelengths in the spectrum are shortened by one thousandth of their value for a body at rest.[9]

---

[9] As in the case of any uniform intermittent motion, as a man walking, the speed will be the number of steps per second multiplied by the length of a step. The same formula applies to light. Light of all colors travels in empty space at the uniform speed of 300 million meters per second (186,000 miles). Thus, for the extreme red light, 375 million million vibrations per second, multiplied by the length of the wave, which is 800 millionths of a millimeter, equals 300 million meters. The waves of violet light vibrate 750 million million times per second, with a wave-length of 400 millionths; multiplying these two numbers gives again the speed of

It is possible to measure the velocity of light with great accuracy, and also to measure, with high precision, the wave-length, as stated above. We cannot count this enormous number of vibrations per second, but derive it by dividing the speed of light by the wave-length.

From the measurements of photographs of stellar spectra made in the last twenty-five years, it has been possible to derive the speed in the line of sight of 2,000 stars. One of the important features of the method is that it gives the result in units like miles per second without requiring any knowledge of the distance of the star, and with an accuracy just as great for a distant star as for a near one. This has given us most important information regarding our stellar system. The speed of the stars is found to range from zero to thirty or forty miles per second, and in exceptional cases, much higher velocities occur. On the average, the yellow stars of class G, like our sun, have a motion in the line of sight of about twelve miles per second. For the white stars of Class A, this average is less—about six miles per second, while for the stars of Class B the velocity is as low as four miles per second. The actual velocities of the stars in space will be twice the velocities here given. Such observations of the speed of the stars furnish us also the velocity of our own sun in space, which was first derived from the apparent motion of the stars across the sky long before the spectroscopic method was applied. It is about twelve miles per second, toward a point lying between the constellations of Lyra and Hercules. If we observe a large enough number of stars in that portion of the sky, we find that we are approaching them with the speed of the sun's motion, and, conversely,

---

light as 300 million meters per second. Thus, if a small boy, representing blue light, is walking with his father, representing red light, with a step half as long, he evidently must take twice as many steps to keep an even speed with his father.

we shall be found to be leaving the stars in the opposite
part of the heavens with the same velocity.

Proper Motions of the Stars.—If stars are moving rapidly
in the line of sight, we should infer that they are also
moving, on the average, with equal rapidity across the
line of sight. This would make the stars drift out of their
places in the constellations, and such a drift was actually
discovered two centuries ago by a comparison of the positions of the stars with the early charts of Ptolemy and
others. Indeed, if Ptolemy were to look upon the heavens
again, he would at once recognize that not a few stars are
out of alignment with their positions in the heavens as he
knew them. Of course, for the average star, this drifting
is very small and can be determined only by precise
measurement with a telescope. The most rapid motion
athwart the sky yet known is that of Barnard's ''run-
away'' star, discovered at the Yerkes Observatory in
1916. This amounts to slightly more than 10″ a year, or
one degree in 350 years.

The researches of Kapteyn have shown that this drifting of the stars is not at random, but that there are two
great streams of stars in our system, meeting at an angle
of about 100°. One of these streams has a trend toward
the constellation Orion, and the other, which includes
fewer stars, is toward the constellation Telescopium. A
careful study by Lewis Boss of the proper motions [10] of the
stars in Taurus, including the Hyades, brought out the fact
that many of these stars are converging toward a common
vanishing point; in other words, they are moving in space
in parallel lines. Since the speed in the line of sight, or
radial velocity, was known for the brighter stars of this
group, it became possible to derive, with a high degree

---

[10] This term is applied to the angular motions of stars across the
sky, and is given in angular measure. A motion of one second per
year, or one degree in thirty-six hundred years, is large.

of precision, the distance of this group, which turned out to be 130 light years. The velocity in space was found to be 29 miles per second, and it was found that if the motion of the stream remained constant, then after 65 million years this whole group would appear as a cluster only one-third of a degree in diameter. This moving cluster, as it may be called, is traveling along with the first of Kapteyn's streams. Another interesting group of stars moving in closely parallel lines is known as the "Bear Family" from the fact that it includes the principal stars, except Alpha, of the constellation Ursa Major, which contains the Great Dipper. The common drift of these stars across the line of sight was noted many years ago, but since the spectroscope has given us the speed in the line of sight it has been possible to determine with some precision the speed in space (eleven miles per second) and direction of motion, which was found to coincide rather closely with that of Kapteyn's second stream. It has been found, further, that some bright stars at widely different points in the sky, such as Sirius, Beta Aurigæ, and Alpha Coronæ, seem from their motion to belong to this group, which is now rather close to us, the five stars in the Dipper being at an average distance of about 70 light years at the present stage of their journey through space.

The Globular Star Clusters.—These extraordinary aggregations of stars occupy a unique place among the celestial objects and are really difficult to correlate with the other celestial masses. The plate made from a photograph taken with the 60-inch reflector of the Mount Wilson Observatory shows the magnificent cluster in Hercules. The plate was exposed for eleven hours, during portions of two nights. This wonderful object is dimly visible to the naked eye, and presents an increasingly numerous constituency as the power of the telescope is enlarged. The stars which form the units of a system like this are brilliant suns, for the most part far surpassing our own sun in luminosity, and it is safe to estimate that there are a million of them in the cluster depicted. The distances of these clusters are so great that the most painstaking measures, both visual and photographic, fail to show any appreciable relative motions during the last thirty years. We believe, however, that each member moves under the combined gravitation of all the others, and that there is no dominant central body which controls the motions of all.

Harlow Shapley, now Director of the Harvard Observatory, studied these clusters very thoroughly while at Mount Wilson and reached, by indirect methods, an estimate of the distances of the clusters. Thus, the cluster in Hercules is placed by him at a distance of about 40,000 light years. The distances of some others were estimated as high as 200,000 light years. This raises the question whether these objects lie within or without our galaxy; if within, then we must greatly extend our former ideas as to the diameter of our own Milky Way system, which was in recent decades supposed to be about 20,000 light years. The late Professor Barnard regarded it to be directly evident from his photographs that certain of the clusters were projected against the stars of the Milky Way. The separation of the individual stars in a globular star cluster, in miles, is very great—of the order of thousands of millions, or even millions of millions of miles. The curious fact is brought out by spectroscopic investigation of clusters, each treated as a whole, that the velocity in the line of sight, which is large, and of the order of ninety miles per second, is in the majority of cases directed toward our sun, as if the clusters were converging upon us. The cluster in Hercules has a motion of approach of about 180 miles per second, or about 1/1000 that of light. It seems at present quite impossible to form any conception as to the sort of body from which these stars have been evolved, but they seem to be a finished product.

The Nebulæ.—These objects are among the most remarkable in the heavens, and are very difficult to explain. We may distinguish four varieties: (1) gaseous nebulæ; (2) diffuse, nongaseous nebulæ; (3) spiral nebulæ, which usually give evidence of very little gaseous content; (4) nonluminous, or dark nebulæ.

(1) *Gaseous Nebulæ.* There are two varieties: (*a*) those which are amorphous, or chaotic in form; and (*b*) the so-called planetary nebulæ, which have a more or less geometric shape, and may be in the form of a ring or rings. These are generally of small angular size as seen from our distance.

The Great Nebula in Orion, depicted in cut facing page 62 is our finest example of a chaotic nebula. It is purely gaseous, as evidenced by the fact that its spectrum consists merely of a few bright lines on a dark background. The gaseous elements represented are hydrogen, helium, and, as the brightest constituent, the gas which we call nebulium, which has not yet been found on the earth. This supposed element is spectroscopically identified by two greenish-blue lines which are not found in the spectra of the ordinary stars.[11]

The great nebula is jeweled with stars which are of an "early" spectral type, such as we have been accustomed to associate with the nebulæ. There is no evidence of interaction between the stars and the nebulæ, despite the fact that some of the stars are spectroscopic binaries revolving about each other rapidly in short periods. If they

---

[11] An hypothesis has been tentatively suggested to account for the nebulium spectrum on the assumption that it is not an element but a combination into a molecule of the atoms of hydrogen, helium and perhaps nitrogen. The chemists' list of ninety-two possible elements ending with uranium has no place for nebulium or coronium.

were immersed in the nebula we should expect such motion to generate a high temperature by friction, with some evident consequences, but they are not found. However, there is other strong evidence that the stars are at the same order of distance as the nebula.

Studies of the nebula with the interferometer and the spectroscope have shown that it is in internal commotion, the speed in the line of sight differing in different parts of the nebula by as much as five or six miles per second. This is a higher speed than that of the molecules of gases at atmospheric pressure on the earth. The Orion nebula is faintly visible to the naked eye, enveloping the middle star, Theta, of the sword handle. Its distance is probably about six hundred light years, and its extent from edge to edge would then be six or eight light years. There are other notable examples of gaseous nebulæ having the most delicate filaments in their structure, but no measurable changes have been found since accurate photographs were first made.

A beautiful example of a planetary nebula is the well-known ring nebula in Lyra. This remarkable object is visible in a small telescope, and like the Orion nebula, is of a purely gaseous nature and of immense size, although of small angular diameter as seen from the earth. From the latest estimate of its distance, we may infer that the space within the ring is ample to contain our solar system, out to Neptune, more than 25,000 times. The spectroscope proves that the ring, as a whole, is approaching us at the rate of twelve miles per second. It would be very interesting to know at what speed the ring is rotating, but it has not yet been possible to determine this with the spectroscope. It would doubtless have been an immense satisfaction to Kant or Laplace to have known of the existence of this ring as a mass of gas. From what has been said, it is evident that if it is evolving from the nebulous into a more substantial system, it would result in something of dimensions far beyond any solar system.

(2) *Diffuse Nebulæ.* There are in certain parts of the sky streaks of nebulous matter which are nongaseous, and which we must regard as sheets of matter, perhaps in the form of dust, which merely reflect the light of some bright stars in their vicinity which illuminate them. Such are the markings in the group of the Pleiades which appear on the photographs like the strokes of a mighty paint-brush. They occur also in different parts of the Milky Way.

(3) *Spiral and Spheroidal Nebulæ.* The spiral nebulæ are extraordinary objects, immensely numerous, presenting themselves at all possible angles with the line of sight. Their spectra are identical with that of a stellar cluster so remote that the stars would not be separately visible. The spectroscope also proves that they are in rotation about an axis perpendicular to their principal plane, and, further, that they are moving in the line of sight at an extraordinary velocity averaging over 350 miles per second for twenty-five of them which have thus far been measured.[12] Incomprehensibly enough, these spirals are, with three exceptions, receding from us as if they were moving away from our stellar system. Their dimensions are incredibly large.

The Great Nebula in Andromeda (opp. p. 254) has a diameter of nearly two degrees, hence it follows that its

---

[12] We should not omit to state that when we use these twenty-five spiral nebulæ as reference points our stellar system is found to have another motion hitherto unmentioned of some 300 miles per second directed toward the present position of the constellation Capricornus. It will be noted that this velocity is the same as that of the average spiral nebula, which is a strong argument for regarding our stellar system as a spiral nebula. The motion of our own sun with respect to the brighter stars of our system is only twelve miles per second. This is like the motion of an individual bee with respect to his fellows in a moving swarm, while the high velocity would represent the velocity of the swarm as a whole.

diameter, in miles, is one-thirtieth of its distance. This nebula, one of the three exceptions, is approaching us with a speed in the line of sight of 186 miles per second, or one thousandth the speed of light. Accordingly, if this speed has been maintained uniformly during a billion years of the history of the earth, the object was then one million light years farther away than now! Other spiral nebulæ have a speed in the line of sight more than four times as great as this, so that it is obvious that many objects of this class are, or have been, at distances of the order of a million light years during the presumable period of development of our earth. The Andromeda nebula is the largest as seen from the earth, and hence probably the nearest. Many spirals are very minute, indicating on the average an enormously great distance. For this and other reasons it may be inferred that as a class spiral nebulæ are at very great distances, and beyond what we ordinarily set as the limits of our galactic system. Estimates of the distance of the Andromeda nebula, however, made by different investigators, vary widely; indeed, from 5,000 light years to half a million.

The period since good photographs of the spirals were first obtained has been too short to yield any results by direct measurement of early and recent plates. Measurements have nevertheless been recently carried out with extreme care by van Maanen on photographs obtained at the Mount Wilson Observatory, over an interval of eleven years. The quantities measured are at the extreme limit of the possibility of detection, but they show an excellent agreement in indicating an outflow of the matter along the arms of the spirals, combined with a radial motion outward from the center. The motion thus derived is so great that a complete revolution would be accomplished in a period of the order of 100,000 years.

Six nebulæ have yielded similar results. If the measurement of photographs taken after a greater lapse of time shall confirm these results, then we shall have to choose the lower limits for the distance of such a nebula as Andromeda (on which measures have not yet shown these angular motions). It is difficult, however, to understand how the supply of matter coming from the inside of a spiral nebula could be maintained if it continues to move out at the rate indicated by van Maanen. In the opinion of the writer, judgment as to the size and distance of these spirals will have to be suspended, perhaps for another decade, until further time has been allowed for these measurements to be repeated and confirmed by independent observers.

The discussion of the so-called theory of the Island Universe, first suggested by Humboldt, has been quite active in recent years and there are many arguments for the view that the spirals represent other galaxies like our own, in corresponding dimensions; but we cannot go into the details here, believing that judgment should be suspended for the present.

(4) *Dark Nebulæ.*—The wonderful photographs of the Milky Way obtained chiefly by Barnard in the last thirty years, reveal extraordinary regions of vacancy, like long lanes or holes in the background of the Milky Way. They gradually came to be recognized as dark markings, and then as dark absorbing objects intervening between us and the background of the Milky Way. They are now accepted by many astronomers as actual dark nebulæ, and 182 of the most notable of these were listed by Barnard in 1919. No doubt the number will be greatly increased as they are more minutely studied. Whether they represent nebulæ which have ceased to be luminous, or such as have not yet become luminous, cannot be decided at present. Possibly they never will become strongly luminous.

UNITY OF THE UNIVERSE

The spectroscope has this important characteristic: the accuracy of its findings are independent of the distance of the object; in other words, the most remote star, provided only that its light is sufficient, can be as rigorously investigated as to its chemical origin as the nearest. With all other astronomical instruments which measure angles, the unavoidable errors of observation expressed in miles increase in proportion with the distance of the object. This great advantage of the spectroscope thus prevents any span of the celestial void from becoming a bar to accuracy in astrophysical research. We can, therefore, compare the constitution of matter throughout the whole depth of the universe, provided only the objects are bright enough for study. This observation on faint, flickering lines of the spectrum barely at the limit of visibility, formerly made with the eye, is now made by photography, the time of exposure being prolonged to make up for the faintness of the less luminous stars.

The striking fact is brought out that with little exception the chemistry of the earth and sun includes that of the whole universe. Hydrogen, the fundamental element in the chemical evolution of matter, gives unfailing evidence of its presence in practically every self-luminous celestial object. Helium, next above hydrogen in the chemical series, is characteristic of the stars of what we call the early type of spectrum (Class B), and doubtless exists in advancing types of spectra of the yellow stars of Class G, like our sun. It was, as a matter of fact, first found in the sun at the eclipse of 1868, as indicated by a slender golden ray in contrast with the dark lines of the other elements. Calcium and magnesium follow, nitrogen, oxygen, silicon and carbon, and then come the other familiar metallic elements: iron, nickel, titanium, etc. Thus, the prin-

cipal elements found in the human body are characteristic of the stars. The dust of which our human structure
is made and to which it returns, even the atoms which form
the tissue of our brain, are representative of the material
of the universe. We can think in terms of the cosmos
because we are of the cosmos.

It might be thought that with the myriad of luminous
bodies populating the heavenly spaces, there would be an
immense variety in quality as well as quantity. We
might have supposed that the different stars in such widely
scattered positions, under such varied conditions of temperature, might show a wholly altered chemical constitution; we might have thought that some great experiments
in the building up of matter were being carried on, resulting in various kinds of material—an entirely different sort
in one star than in another. This, however, is not the fact.
*The universe is composed of essentially the same atoms
wherever found.* It is a cosmos.

Stellar Cataclysms

The course of a star's evolution does not always run true.
Events occur with startling suddenness which may in a day
increase the splendor of a sun by many thousandfold.
Such occurrences are not rare; indeed, two such temporary
stars, Nova Persei of 1901 and Nova Aquilæ of 1918, surpassed the first magnitude. The latter, as proven by subsequent examinations of photographs, had led an unevent-
ful existence during the twenty-five years since it was first
recorded on the sensitive film, fluctuating slightly around
the tenth magnitude. On the morning of June 8, the day
of the American total solar eclipse, it was caught on a
Harvard plate as just bright enough to be visible to the
naked eye; at dusk on the same evening, it rivaled Altair,
of the first magnitude, and on the following evening, Sun-
day, June 9, it was the brightest star visible in the northern

heavens, out-shining Vega. If that sun were attended by a group of circulating planets, then, for any possible inhabitants, it was

> Dies iræ, dies illa
> Solvet sæclum in favilla.

After an ephemeral period of splendor, the star waned; its fluctuations grew fainter until, after about six months, it disappeared from the sight of the naked eye. Then, for a period, it exhibited the spectrum of a gaseous nebula, and now is again back at its former brightness, one hundred times too faint for the unaided eye to see. No longer is it outstanding among its millions of fellow suns of like magnitude.

It was such an object which flashed out in 134 B. C., and caused Hipparchus to catalogue the stars. The most brilliant Nova yet recorded, that of 1572, put a dramatic element into the early observations of Tycho. Kepler's star of 1604 was not quite so bright, although it surpassed all of the other stars of the heavens and was visible for over two years.

The causes of the outbursts of such temporary stars are not known, but the spectral histories of those which have been visible in the past thirty years have been so similar as to show that it is a regular procedure, occurring under uniform physical law, not representing a breaking down of law. Although many astronomers still believe that a collision in one form or another is responsible, to the writer it seems more probable that their sudden brilliance is due to the release of atomic energies which are doubtless adequate for the purpose, although we cannot understand how they come into play.

COSMOGONY

A statement of the theory of the origin of the solar system, or indeed of the stellar system, may perhaps be expected by the reader whose patience has led him thus far in this chapter. The development of a cosmogony has been the sport of philosophers, particularly in the early days before there was the present insistence of basing theory upon facts of observation. We must disappoint the reader because the discoveries of the last two decades have tended to discredit all of the systems of cosmogony, however long held or persistently defended. There are flagrant contradictions of reasonable interpretation of facts in every theory of cosmogony thus far presented. As a matter of fact, the most difficult part of the philosophy of the universe, both mathematically and physically, is to account for the origin of a solar system, that is, for a retinue of planets traveling about the central star, these planets themselves attended by subordinate satellites.

It would certainly be a serious omission if we were to leave out from our statement a recapitulation of some of the facts of observation which compel us to believe in an orderly development of the solar system, although we are unable to give any satisfactory theory of its beginning.

The sun is a great sphere chiefly, if not wholly, gaseous, maintaining its output of radiation without sensible diminution. The sun turns on its axis in a direction which is called west to east. The earth and a thousand other known planetary bodies move about the sun in the same direction and in planes not greatly inclined to that of the earth's orbit, the ecliptic. The principal planets turn on their axes in the same direction, and these axes are approximately parallel to each other and to that of the sun.

The major planets have a considerable number of satellites whose direction and plane of motion correspond, with certain exceptions, to those of the rest of the objects in the system: nine (perhaps ten) have been found to surround Saturn, and nine have been detected about Jupiter. The rings of Saturn consist of immense numbers of small satellites revolving about the planet under the law of gravitation, just as do the larger moons.

The planets seem to exhibit a certain relation between their present condition and their distance from the sun. Thus, the earth and Mars, Jupiter and Saturn, Venus and Mercury, and finally Uranus and Neptune seem to have various characteristics in pairs.

There is also a certain orderly relation of the distance of the planets from the sun which has thus far been expressible only as an arithmetical series, not as a law. It fails for Neptune, but does not for a planet at the average distance of the asteroids. There is reasonable evidence that the group of asteroids represent a planet which would have been found between Mars and Jupiter, but which perhaps suffered an explosion, or a series of disruptions, in the course of its development.

Some exceptions to the regularities noted above are the probable retrograde rotation of Uranus and a similar motion of all of its satellites, and the very short period of revolution of the inner satellite of Mars, which describes its circuit about the planet in about one-third of the time that the planet turns on its axis. The other exceptions are the retrograde motions of Phœbe, the ninth satellite of Saturn, and of the eighth and ninth satellites of Jupiter. The rather uniform behavior of these exceptional satellites strongly suggests that they are not actual exceptions to, but examples of, a law which is not yet understood.

From the study of such of the above correlations as were known by observation in his time, Laplace in 1796 rather diffidently and incidentally brought out his *Système du Monde* The theory advanced in it later came to be known as the Nebular Hypothesis of Laplace. Nearly half a century before, a somewhat similar theory had been advanced by Immanuel Kant, and still earlier by Thomas Wright, but Laplace does not seem to have known of it. The theory postulated a gaseous nebula rotating on an axis and contracting, accompanied by a tendency to flatten into the form of a disk. It was thought that when the centrifugal force from rotation equaled the gravitational attraction, a ring would be left behind which would perhaps later develop into a planet, possibly with its own system of satellites evolving in a similar manner. The sun would be the remaining part of the primitive nebula, the planets among them, of course, the earth, having been formed from its outer portion.

The work of H. von Helmholz (1854), in showing that the sun's heat might be maintained by a slow contraction of its sphere, fitted admirably with this theory. It was not an induction; no one knew in Laplace's time that the nebulæ were actually gaseous, although this was guessed from external appearances. The elder Herschel had observed a nebulous star in Taurus which had set him to thinking about these matters, and thus Laplace received some information about the possibility of the existence of such objects as was evidently referred to in later editions of his work. It was not until 1864 that Huggins, on first directing the spectroscope to a nebula, was surprised to find that its light consisted only of a few bright lines, a discovery which would have been also of extraordinary interest to Laplace and the elder Herschel. Gaseous nebulæ were thus proven to exist, and were not imaginary. Furthermore, as the spectra of the different types of stars were examined, it became possible to show that there was an unmistakable sequence in the evolution of the stars. No terrestrial race could expect to be so long perpetuated on its planet that its seers might watch the transition from one stage to the next; but from examining the sequence, there could be no doubt that, given time enough, any star like our sun would doubtless go through an orderly succession of changes, physical and chemical, for each stage of which sufficient examples are present in the heavens. Nebulous stars were also found to be tolerably numerous, and the number of so-called planetary nebulæ was greatly enlarged by the discoveries with the spectroscope. This evidence, much of which was outside the original scope of the hypothesis of Laplace as to the origin of the solar system, tended greatly to support the broadened nebular hypothesis.

But mathematical analysis has shown that this theory is not correct for our solar system for which the moment of momentum can be computed, and is inadequate even though our sun has contracted from a diameter extending out to the orbit of Neptune. Further than this, we are beginning more fully to realize that the gaseous nebulæ with which we are familiar are all too large to produce so small a thing as our solar system. We mentioned that a recent estimate of the distance of the ring nebula of Lyra makes an area within the ring large enough to hold thousands of solar systems like ours. The great gaseous nebula in Orion is big enough to supply material for thousands of suns instead of a single one. Again, a theory has recently been advanced, with some considerable evidence in its support, that the nebulæ, even the gaseous ones, borrow their light from certain stars in their vicinity, and give it out in altered frequency of vibration—in other words, they take the light of a star, and convert it into that of a nebula. We must say that we cannot regard this evidence sufficient to convict the nebulous matter of dishonesty in celestial places in purloining light of stars and emitting it as it were under false pretenses. But there is a growing doubt as to the necessary evolutionary relation between stars and nebulæ. At a certain stage in the decline of a nova, or temporary star, its light becomes that of a nebula for a time, and later returns to that of a star. Local and admittedly exceptional conditions may then do in a few weeks what would be supposed to require æons of time in any nebular hypothesis.

The effect of tidal reactions has been adduced in support of the nebular hypothesis as well as other phenomena, and with advantage, but all theories which have to depend upon the casual close approach of two suns for the tidal disruptions supposed to be necessary to produce a train of planets and satellites, demand lengths of time which seem inordinate in view of the great evidence of multiple systems. Of ordinary visual double stars, some 15,000 have already been discovered, or one in about eighteen of the stars examined for duplicity with powerful telescopes. We have not had space to discuss previously the great number of double stars that have been discovered with the spectroscope, which were too close to be detected by the ordinary telescopic means. From the results of the application of the spectroscopic method, where the approach of one member of a pair and the recession of the other member will show lines alternately shifting back and forth or doubling, we find that among the brighter stars nearly every other one is double. Many well-known stars are found to be even triple or quadruple systems, and there is every evidence that such systems are of a natural order. Observation would thus seem to indicate that fission must be a very common characteristic of evolved suns, although the mathematical demonstration of the mode of partition may be very difficult.

The meteoritic hypothesis of Lockyer, which endeavored to develop the solar system from swarms of meteorites, was based on inferences derived from insufficiently accurate spectroscopic observations. It was correct in assuming that a star in its evolution ought to have an ascending as well as a descending scale of development of temperature; that is, it should pass through various spectral classes as it proceeds toward its point of maximum temperature and brilliance, as well as the reverse order as it declines from its highest stage of thermal development. A star should thus twice pass through the stage of redness, once on the rise and once on the decline. H. N. Russell has shown that the giant stage of a star like Betelgeuse is on the rising side, and the dwarf stage on the declining side of the arch of the temperature curve.

The planetesimal hypothesis of Chamberlin and Moulton, which builds up the planets and satellites from meteoric dust resulting from a partial disruption of two suns which have accidentally come close enough to each other, is doubtless mathematically sound. It is very unfortunate, however, in assuming that a spiral nebula would result from such an approach of two stars, and be an intermediate stage in the development of a system of planets. We now know that the spiral nebulæ are objects of an entirely different order, adequate to form systems of thousands or millions of stars instead of a paltry group of inconsequential planets. Any theory of cosmogony which starts with a finished product, a sun, appears illogical and unsatisfactory, but we must, of course, distinguish between a theory which strives to account only for the presence of our sun and its satellites, and a general theory which endeavors to explain the evolution of the stars themselves.

One of the latest investigators in the field of cosmogony, J. H. Jeans, of Cambridge, England, establishes a mathematical basis for a theory which closely follows that of Laplace; but he finds that the scale of mass of his lenticular body of gas makes a great distinction in the outcome as mathematically developed. If the nebula has a thousand or a million times the mass of our sun, then the development will be into something like a spiral nebula, from the knots of which many stars will develop. If it is small, and has the mass of a single sun, then the procedure is greatly altered. His conclusions were not accepted by all students of cosmogony, and he frankly says, at the close of his Halley Lecture, delivered May 23, 1922:

". . . If the whole aim of cosmogony were to discover the origin of our own system, our labour would have been in vain. But an intelligent cosmogony will have a more objective aim, and the cosmogonist will be concerned to gain a knowledge of the origins of the stars as a whole rather than of the genealogical tree of our own particular planet. Judged by the wider standard, Laplace's conception has been amazingly fruitful. It would hardly be too much to say that it has either revealed or given a valuable clue to the origin of every normal formation in the sky, with the single exception of that of the solar system which it set out to seek."

Jeans even questions whether a solar system like ours is the normal evolutionary result.

Twenty years ago Alfred Russell Wallace, eminent as a contemporary of Charles Darwin in the field of biological evolution, published a large work entitled "Man's Place in the Universe," in which he made our star an exceptional one at the center of the galaxy. His arguments, though ably presented, were not convincing. We should not wish to leave with the reader the impression that these views of even so eminent an investigator as Professor Jeans have been widely accepted by astronomers. It seems more reasonable to think that the difficulties of accounting for a solar system are due to the inadequacy of our present mathematical analysis rather than to the limitations of nature itself.

In this chapter we have presented some of the facts now known in regard to the stellar universe and to that minute portion of it called the solar system. There are necessarily many gaps in the theories which endeavor to account for these facts. But the progress of science is not made or illuminated by dogmatic assertions in respect to theories, and the laborious process of testing hypotheses by observation, and then proceeding to new theories will be continued as long as the mind of man concerns itself with such a stupendous reality as the Cosmos.

CHAPTER V

THE MAKING OF OUR EARTH

By Edward B. Mathews [1]

THE outstanding contributions of Geology are a knowledge of the earth as it is, and the addition of several
fundamental ideas to our thinking. Among the
latter are the increased conception of time since the earth
came into being; the efficiency of small forces, if given
time, to accomplish immense results; and the persistence
of forces, processes and conditions on the earth without
fundamental modification during the course of geological
time.

The views now held of the earth, its history since its
beginning, and the processes by which it has come to its
present state, are based upon facts established only after
the study of confirmatory facts observed in many parts
of the world, and upon theories or hypotheses which have
been accepted only after years of critical consideration
and often controversial debates lasting frequently for centuries. The facts are so numerous, so diverse in character
and so widely scattered over the face of the earth that a
single student must accept the statements of accredited
witnesses for all but those which may be studied in a
single lifetime in the limited area open to his observation.
The forces involved in the processes and the time through
which they have acted are so great that like conditions
can seldom be reproduced experimentally. The geologist
must therefore investigate in humility and amazement that

[1] Professor of Geology, Johns Hopkins University.

his carefully classified observations yield such stupendous conclusions.

Probably the easiest way to comprehend the simplicity of the steps of progress and the far-reaching conclusions attained would be through a glimpse at the views held before the Dark Ages and a tracing of the development of knowledge since the Renaissance. In such a rapid survey it should be borne in mind that the Greek natural philosophers gained many facts and correct views concerning the earth. Since these were not generally accepted at the time they were overlooked by the Romans and lost during the Dark Ages to be rediscovered during the Renaissance.

The knowledge of the geography of the earth up to the third century of the Christian era was based almost entirely on the works of Eratosthenes, who lived 750 B. C., and those of Strabo and Ptolemy which, transmitted by the Arabs, became the main geographical authority at the beginning of the Renaissance. Some of the Greeks had recognized the daily rotation of the earth and its annual revolution about the sun, but the ancient world generally thought of the earth as a sphere or disk in a starry dome chiefly covered with water on which the inhabited portion was an elliptical area, divided into three continents, extending north and south from the British Isles to Arabia and Abyssinia, and east and west from the Atlantic to Ceylon and probably, by rumor, to China. Even the most enlightened knew little or nothing of the geological history of the earth and very few of the natural laws on which the science of Geology is based. They pictured the earth as the center of the universe, the lands about the Mediterranean as the center of terrestrial features, and the people of this region as the center of human activities. Such conceptions saturated the daily language of the common people and colored the speculations of philosophers and theologians. Religious teachings were naturally given in such a false physical setting since in no other way could they be presented by teachers to people who held such conceptions of the earth.

When, however, the sober-sided western mind insisted that the physical frame must be accepted in all its details great harm was done by forcing into opposition the traditional tenets of the Church to the increasing knowledge of physical phenomena. Among the sad pictures of human history are those of banishment by the Church of its most brilliant and idealistic sons because they would not accept as true scientific beliefs which the Church has later discarded.

Before there could be sound geological interpretations, all these ideas of a little static earth at the center of the universe must necessarily be enlarged and modified. Until the world accepted the views of Copernicus, that the earth moved around the sun, and those of Galileo that the earth was smaller than the sun, and revolved on its axis, geological students struggled in an adverse environment. Philosophical guesses, some of which were ultimately proved true, were made but, as Zittel says:

"Not a single writer of the ancient world showed any interest in the firm earth crust, not one observer gave a thought to the composition of the rocks. Not the most acute thinker of those cultured peoples had even a shadowy premonition of the value that might appertain to fossils as witnesses of a sequence of events in the history of the earth. None suggested that our planet might have passed through a succession of changes before attaining its present physical condition and configuration; still less, that particular phases in the history of change might be deciphered from the character and superposition of the rocks. The evolution of the earth and its denizens, which is at the present day the great problem of geological and biological research, played no part in the literature of antiquity."[2]

DISCOVERY OF THE EARTH'S HISTORICAL RECORD

The fifteenth and sixteenth centuries were periods of phenomenal outreachings of the human mind when the revolutionary suggestions made were too great for the average man to accept without careful consideration and long deliberation. Out of these centuries and those which followed came finally the acceptance of an enlarged conception of the solar system which placed the earth in its position as one of the smaller planets moving around the sun. During the same period began the series of spectacular voyages of discovery which placed the Mediterranean and its inhabitants as but a part of a more extensive world inhabited by men not very different from the more civilized heirs of classical learning. There yet remained a lack of any clear appreciation that the earth itself had a historical record of its own which could be deciphered.

The Greeks from Xenophanes of Colophon (B. C. 146 †) to Aristotle had noted shell- and leaflike forms in the rocks far from the sea and had mused on what they meant, and on how they were formed. These speculations continued long after another of the fifteenth century intellectual giants, the artist Leonardo da Vinci, had correctly shown how such fossils were the remains of once-living organisms which had been buried in the muds now hardened into solid rock. This explanation was too simple for belief in a day when the earth was still thought of as created in six days and peopled by supernatural fiat. It was much easier to assume some mystic "plastic force" or that the fossils were due to some occult action of the stars, or even that they were the buried unsuccessful efforts of the Creator

---

[2] Zittel, *History of Geology and Paleontology*, Ogilvie-Gordon translation, p. 11.

before he fashioned the present inhabitants of the earth. Later the belief that fossils really represented the shells and bones of once-living animals was accepted, but their present occurrence in the rocks in the mountains far from the sea was thought to be due to the Noahian deluge rather than to changes of the relative level of land and sea. This controversy raged for full three centuries and echoes of it may still be heard in quiet corners removed from the currents of scientific thought.

While these controversies were going on students were gathering fossils, and facts concerned with their occurrence. Among these perhaps the most significant was the relation between the fossils and the rocks in which they occurred. It became thoroughly established that certain forms were found only in certain beds and thus that different beds were characterized by different fossils. If one could tell the relative age of the beds containing these fossils one might find a record of the life on the earth at different times.

Nickolas Steno (1638-1687), a Dane, teaching at Padua, supplied a key to the problem which has been amplified with later knowledge. The reasoning under his "laws" is simple: any definite layer of deposits must have been formed on a solid base which must have consolidated before the fresh deposit was laid down. Such deposits might be widespread or local. Since the deposit must have fallen through some lighter medium, such as water or air, the older layers must be beneath the younger. Finally if the originally horizontal beds have been uplifted by earth movements then in areas of disturbance there may be evidence of relative ages by the degree of movement manifest. With the orderly sequence of deposition and other criteria developed by subsequent investigators, it became possible to determine the relative ages of rocks and their contained fossils.

Finally, William Smith, the Father of English Geology, by innumerable observations established the fact that similar successions of rocks and fossils could be recognized at different localities and that such different occurrences could be connected by tracing prominent ledge-forming layers, or even by the finding of characteristic fossils in the intervening area. This work, with that of Cuvier and Brongniart in France, demonstrated the possibility of mapping individual formations characterized by specific fossils and the reading of geological history from such geological maps.

THE STORY OF THE ROCKS

The years during which it was slowly being realized that there was a history of the earth and that parts of it might be read from leaves of stone bearing characters formed by the life of the long-ago, were not years of listless waiting. Geologists had been studying the bits of the record exposed here and there and collecting the fossils characterizing the different beds. Paleontologists had been gathering these fossils into museum collections, studying their affinities and describing their features so that similar forms might be recognized at distant points and called by the same names. When the conclusions of Brongniart, Cuvier, and William Smith became known these organic relics characterizing past ages became objects of widespread interest. A knowledge of fossils was no longer the hobby of a dilettante but the goal of highly endowed men who gathered themselves into groups and societies to further the accumulation of facts from which to construct a history of the earth. That work is still going on with greater and greater refinement but enough has been deciphered to make a fairly connected story.

Such a story cannot be narrated in the units of time used by historians of human activities. It has been neces-

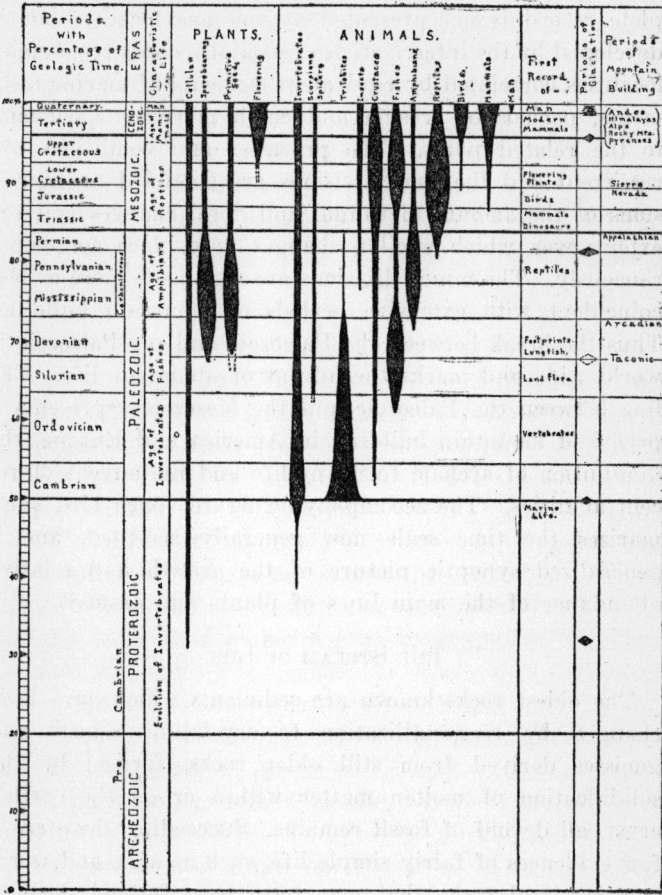

Periods, With Percentage of Geologic Time.	ERAS	Characteristic Life	PLANTS.				ANIMALS.											First Record of –	Periods of Glaciation	Periods of Mountain Building
			Cellular	Spore-bearing	Palm-like Cycads	Flowering	Invertebrates	Scorpions Spiders	Trilobites	Corals	Crustacea	Fishes	Amphibians	Reptiles	Birds	Mammals	Man			
Quaternary	CENOZOIC	Man / Age of Mammals																Man		Andes Himalayas Alps Rocky Mts Pyrenees
Tertiary																		Modern Mammals		
Upper Cretaceous	MESOZOIC	Age of Reptiles																Flowering Plants		Sierra Nevada
Lower Cretaceous																				
Jurassic																		Birds		
Triassic																		Mammals Dinosaurs		
Permian	PALEOZOIC / Carboniferous	Age of Amphibians																		Appalachian
Pennsylvanian																		Reptiles		
Mississippian																				
Devonian		Age of Fishes																Footprints Lungfish		Arcadian
Silurian																		Land Flora		Taconic
Ordovician		Age of Invertebrates																Vertebrates		
Cambrian																		Marine Life		
Cambrian	PROTEROZOIC	Evolution of Invertebrates																		
Pre Cambrian																				
	ARCHEOZOIC																			

CONSPECTUS OF GEOLOGICAL HISTORY.

sary for geologists to construct a time scale in larger units
which do not imply strict synchroneity, but equivalence in
development of earth structures and fossil life. The complete record is not presented at any one locality, but is
developed by the interrelation of partial records from many
localities correlated by equivalent elements of marine sediments, periods of erosion and earth movements common
to the related parts. The processes and conditions are
recurrent and the time divisions are classified as expressions of the amount of faunal and floral changes and the
extent over which similar changes took place contemporaneously. The minor divisions are also made more or less
coincident with extensive periods of mountain building.
Thus the break between the Paleozoic and pre-Paleozoic is
world wide and marks the advent of abundant life. The
line between the Paleozoic and the Mesozoic represents a
period of mountain building in America and Europe, the
diminution of archaic forms of life and the advent of recent affinities. The accompanying figure (page 116) sum-
marizes the time scale now generally accepted, and a
generalized synoptic picture of the growth and relative
abundance of the main lines of plants and animals.

THE STREAM OF LIFE

The oldest rocks known are sediments which have been
changed by recrystallization to crystalline schists and
gneisses derived from still older rocks, formed by the
solidification of molten matter within or on the earth's
crust, all devoid of fossil remains. Succeeding these are a
few evidences of fairly simple life, such as algæ and worm
burrows, and rocks that may have been formed through
the life processes of plants and animals which have left
no other trace of their existence.

When and why animal forms acquired hard shells capable of preservation is unknown, but at the beginning of

the Paleozoic, in the lowest beds of Cambrian age, such forms are present in abundance. Fully one thousand species are now known from the rocks belonging to Cambrian time. Most of these were brachiopods and trilobites with a few sponges and corals. This sudden expansion in the records of life means that some pages of the story are still missing. There is no positive evidence that land plants existed at this time, but such as may have existed would hardly leave any trace. No fragments of the more resistant woody plants have been found below the younger Ordovician rocks.

Many of the types of mollusca were represented before the end of the Cambrian, but they were relatively unimportant until much later.

The oldest evidences of fishes, early vertebrates, are found in the next succeeding period, the Ordovician, though their ancestors may have lived earlier. These were followed in the next period, the Silurian, by evidences of the oldest air-breathing animals, the scorpions. Though the scorpions of to-day live on land and feed on insects, spiders, and other small forms, the first representatives probably lived in shallow water or along the shore where they could find food. The scorpions and the huge crablike eurypterids were among the first animals to travel up the estuaries into the fresh-water rivers.

From the beginning of Paleozoic time the low-lying hills may have been green, for Silurian rocks show some evidence of land plants. These, however, are rare and poorly preserved as they lack woody tissue which is first found in abundance in the rocks of the Devonian. Subsequently the plants became larger, more abundant, and varied until the great development recorded in the Carboniferous rocks from which more than 3,000 species have been described. Probably more than half of these are spore-bearing and seedless, the remainder being seed-bearing ferns and conifers and possibly a few true flowering plants.

Prior to the covering of the land with plants there was no suitable habitat for land animals. The rocks from the Carboniferous onward, however, show an ever increasing evidence of vertebrates which lived on the land.

The last fifth of geological history includes the growth of the reptiles of huge bodies and little brains, the development of the mammals and finally man himself, the first animal, though of relatively insignificant physical strength, that has been able to utilize consciously the forces of nature for his own enrichment.

This history in the rocks brings to light many interesting facts concerning the unfolding of life, the length of geologic time, the persistence of geologic processes and the efficiency of small forces. The *unfolding of life* has not been haphazard, but in accord with principles, the phrasing of many of which are still beyond the grasp of man. When the fossils are compared in their chronological sequence it becomes evident that "the living population of the globe has undergone almost continuous change, old forms becoming extinct, and newer, more specialized forms taking their place, the change being in general from lower to higher." While a few forms, like Lingula, living in simple environments have persisted practically unchanged from the Cambrian to the present, the general rule has been that specializations to meet complex environments have reduced the power to make the anatomical and functional adjustments necessary to withstand changes in physical conditions, and so caused extinction of succeeding groups. As in art over-decoration is an evidence of decadence so among fossils development of ornamentation foreshadows decline and possible extinction. When the material is sufficiently abundant, the intervening differences are bound to be slight. Waagen (1869), comparing the ammonites from successive layers of Jurassic rocks, was able to observe the minute and inconspicuous changes in form which resulted in the gradual emergence of successive new species. Geologists, the world over, investigating other animal groups of different geological ages, have established with varying degrees of completeness similar series of evolutionary gradations. To every geologist of repute evolution of organisms during geologic time is an indisputable fact. Whatever may be the validity of any theory or hypothesis offered to explain its cause or the mode of its operation, the evolution of organic forms is clearly set forth in the history of life written in the record of the rocks. Here we have the original documents arranged in chronological order. The explanation of the changes is the problem of the biologist rather than the geologist.

The *lapse of time* during which the rocks have been laid down is far greater than once supposed. How great in human units of time has not been determined. Many methods have been used to solve this interesting problem, but all contain unknown variables and the results are merely statements of the relative duration of geologic epochs, or estimates of the probable age of the earth. The relative proportions are shown in the accompanying diagram; the absolute duration is in millions, perhaps trillions, of years.[3]

---

[3] The uncertainty at present is largely due to a recent recognition that radioactivity affects the former conclusions of physicists as to the duration of the earth as a planetary body. They now suggest that the earth may have been much as it is to-day for one or more thousands of million years. The geologists, while formerly criticized for excessive estimates of 100 million years, are not yet ready to accept such figures. An intermediate position suggests about 250 million years for the period from early Archean time to the present.

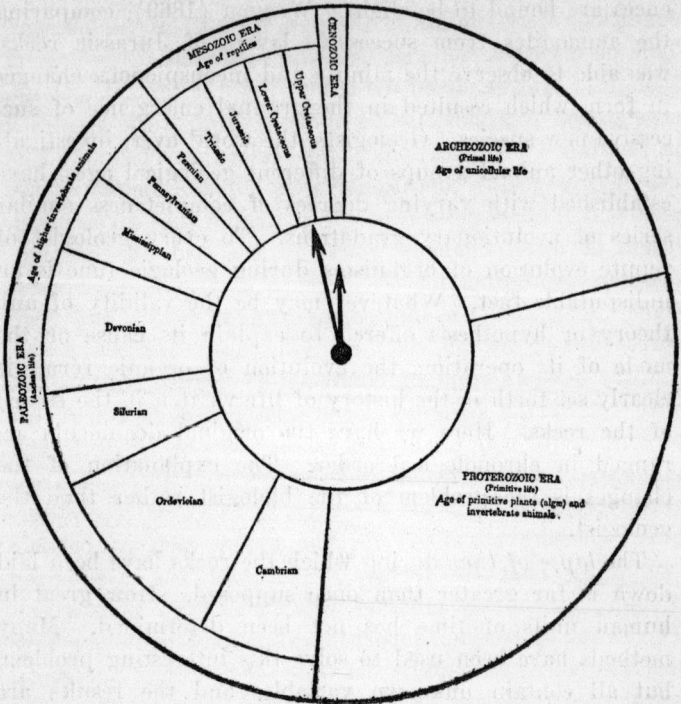

GEOLOGIC TIME CLOCK.
(From Berry, "Geologic Evidence of Evolution,"
Scientific Monthly, xv, 2.)

Prior to the unraveling of the geologic record there was little thought of an earth much older than the race of man and the duration of this period was thought of as only a few thousands of years. The ancient conceptions of time were even less accurate than those of the size of the earth's surface.

The presence in the rocks of ripple marks left by the tide, the impressions of worm trails and the tracks of animals like those found in estuaries to-day, suggested that conditions were similar in the past to those of the present. This observation changed the whole current of geological thought. Hutton, from his study of the hills of Scotland, crystallized this view in his epoch-making work amplified in Playfair's "Illustrations of the Huttonian Theory." The use of present-day conditions to explain the history of the past and the recognition of the principle that an accumulation of small changes would in the end produce stupendous results became the basis of geological interpretations. The idea was not entirely new, for it had been foreshadowed by philosophers like Kant and Descartes, but the accumulation of confirmatory evidence presented with convincing logic by Lyell established the principle of uniformity which now lies as the basis of all geological thought. This conception and the new conception of great geological time were the necessary background for subsequent theories advanced for the explanation of organic evolution. They also formed the basis for the explanation of many previously misinterpreted geological phenomena. The ratios of erosion, deposition and other processes have not remained absolutely constant throughout geological history for there are evidences of accelerated and retarded activity, but during the latter half of this long lapse of time the changes have taken place as imperceptibly as they do to-day.

EFFICIENCY OF SMALL FORCES

The sane explanation of past events of stupendous scope by the application of familiar processes working for long periods of time involves a new conception of the efficiency of small forces causing small increments of change to produce immense results. Before Hutton presented a picture of a calm Nature steadily working little by little during long periods of time, the philosophers were wont to call upon great forces working catastrophically. Such an unnecessary invocation of supernatural forces was natural in the days of mythology and the childhood of the race and is still frequent in the imagery of children. Unfortunately it still persists among many adults when they offer explanations of earth phenonema with which they are familiar. To-day, even among those educated in other lines, a newspaper report of earthquakes will produce a surprising crop of catastrophic interpretations of their origin, as it is much easier to invoke some great unknown force than to conceive of the adjustments of stresses and strains which have been accumulating little by little until they equaled or surpassed the strength of materials in the earth's crust. This idea of the efficiency of small increments of change is the basis of all hopes of success "by keeping everlastingly at it," whether the success is sought in business, education or moral improvement, but the layman hesitates to apply it in the field where it was first recognized to be true.

Glacial periods may serve as an illustration of small forces producing outstanding results. If but a few flakes of snow should fall each winter in excess of the melting of the summer, the earth would at length be covered with ice. Periods of excessive ice cover have been recorded in the pre-Cambrian, the Permian, the Eocene and the Quaternary or Glacial Epoch. The proposed explanations of glacial climates under which excessive snow is deposited have been many. No single cause is accepted, as these climates are now thought to be due to combinations of several factors. It is not thought by geologists generally that these periods of cold are due to wanderings of the Poles, to decrease in the carbon dioxide of the atmosphere, or to atmospheric blankets of volcanic dust though these causes might be invoked if other facts were favorable. It seems more probable that glacial climates are due to the combined influence of changes in solar radiation and to changes in the surface of the earth during and following periods of mountain elevation affecting the extent and distribution of inland seas, the movement of water circulation in these seas, and the concentration of precipitation on uplifted slopes. The cause of warmer interglacial climate such as that, perhaps, at present is probably due to a genial oscillation in the solar radiation.

ORIGIN OF THE EARTH

Prior to the Renaissance, the common people and philosophers alike had little conception of how the earth was formed. When attempts to explain its origin were made, few facts were available to limit their flights of imagination which usually involved some Supernatural Power who worked rapidly, producing in a few steps the earth as it was then known. To-day the terrestrial sphere, with its present configuration, is regarded as but an expression of the contemporary stage of development of one of the lesser planets of our solar system, which is, in turn, only one, perhaps a minor one, of unnumbered similar systems. The explanation of the origin of these systems and even of the earth itself belongs outside the field of geology, though geologists are forced to picture the condition of its surface when it first became the abode of living organisms, and to explain the geological history during the long eons prior thereto.[4]

---

[4] Kant thought of matter diffused through space which subsequently became segregated through the mutual attraction of its particles into swarms which became condensed to planets. The initial heat would depend upon the mass of the growing body and the rate of condensation. Laplace, fifty years later, pictured the planets and their satellites as derived from an attenuated gas at high temperature. The material of the earth first formed gaseous rings, then a liquid globule, and finally a solid sphere. The earth was formerly at a high temperature and is now a cooling body. Chamberlin's planetesimal hypothesis presents the picture of an original central body or sun disrupted by tidal distortion, throwing

The geological phenomena indicate that as far back as they can be interpreted the processes of erosion and deposition, with rivers, estuaries, tides and seashore, were the same as to-day. The climate also was similar, evidences of warm, moderate and even glacial climates have been recognized in rocks of widely different ages from the oldest to the youngest. There were rain and sunshine, highlands and lowlands, oceans and continental areas as there are to-day. Whether these were distributed over the earth's geoid as they are at the present is doubtful. Many geologists believe that during the very earliest geological times the earth was somewhat (200 to 400 miles) larger than at present and that there has been a progressive shrinkage causing foldings of the upper crust, the trend of the folds being determined perhaps by already existing continental and oceanic areas. The cause of the shrinkage and crustal shortening was formerly ascribed to the shriveling of the outer crust of a cooling earth, but with our advancing knowledge of physical chemistry, especially of crystallization and radioactivity, the shrinkage is ascribed by many to recrystallization into denser molecules going on within the depths of the earth.

Although the positions of the continents and ocean basins have shown little change, the outlines have varied greatly during geologic time due to the oscillation of the coast line and the expansion and contraction of shallow continental seas, like the Baltic of to-day.

From the geological affinities of fossils, similarities in sedimentary series and geological foldings it has been shown that the waters of inland seas have frequently

---

off matter in spiral nebulæ and spiral knots, one of which became the earth. This gradually grew by accretion of particles or planetesimals. The initial temperature was not high at the surface, but increased internally by condensation and molecular rearrangements.

covered large portions of the present American continents. According to current paleogeographic maps, no part of the North American continent has remained as land continuously throughout Paleozoic time except, possibly, part of the geologically little-known Labrador peninsula. At times during the same eons there were probably land or shallow-water connections with Europe, Asia and possibly Africa.

The vertical movements due to mountain-making, regional uplifts and subsidences have been generally accepted, but little is known regarding horizontal movements. Slight horizontal changes are known to be taking place in California, but the theory of a widespread horizontal drifting of continents, as recently proposed by Wegner, has not secured general acceptance.

THE EARTH'S INTERIOR

Man cannot examine directly the interior of the earth. The deepest mines scarcely penetrate one mile from the average surface curve. Little can be proved concerning this interior beyond the fact that it is nearly twice as dense as the rocks at the surface and as rigid as glass or steel. There is, moreover, near the surface an increase of heat as one penetrates the crust of the earth, and this increase probably continues to great depths on account of the enormous pressure exerted by the overlying rocks. Many phenomena indicate that the composition of the earth becomes progressively more basic and less siliceous with depth, and this separation into concentric layers of differing composition, especially near the surface, has apparently taken place in accordance with well-established physico-chemical laws.

That the popular assumption of a liquid interior covered by a thin crust is not true is deducible from the manner of transmission of earthquake shocks, the rigidity of the earth under tidal strains and many other phenomena.

The traditional explanation of volcanoes has been that they are "safety valves" through which highly heated portions of the interior, though solid, have been forced to the surface along lines of weakness where they have become liquid through relief of pressure before the temperature has been lowered below the melting point. According to this theory the heat involved has come from the interior, as a remnant of the original heat of the earth, or subsequently developed by the crushing and compressing of the earth's crust. The accompanying gases, according to this view, were part of the original content of the earth or were in some way derived by percolation from nearby seas. Quite recently the experiments made in the lava lake at Kilauea show that the temperature of the lava does not increase with depth and that it is hottest at the surface. Moreover, the lava has been found to be more viscous and richer in gases beneath the surface. From these and many other observations it is now inferred that the great heat of the lavas is due to chemical reactions in escaping gases, that the volcanoes are essentially local, crustal phenomena unconnected by conduits filled with molten matter with a deep-seated, highly-heated interior.

This growth in knowledge concerning the earth and how it is formed furnishes a background for philosophical conceptions just as the more meager and inaccurate knowledge did among the ancients. The realization that the same processes of erosion and deposition, the same variations in climatic conditions from benign to glacial, and the same physical and chemical reactions, have taken place as they do to-day under similar conditions during the millions of years whose history has been deciphered makes any easy assumption of highly different conditions presumptively false. The nonchalant postulation of breaks in the uniform action of these processes to meet a personal prejudice or some fancied need in biblical interpretation grates upon the senses of those who are filled with awe at the marvelous mechanisms which have remained in so delicate adjustment for eons. Fundamental truths clothed with such accessories fail to make the impression on the modern mind which are their due.

The orderly development of life step by step through the ages renders impossible any credence in philosophies involving any theory of successive special creations, while the firmly established chronology of the rocks in which the fossils are formed precludes belief in any sort of synchronous special creation of the forms of life we now know.

The shift in view with respect to the position of man is at once striking and thought-provoking. Only within the last century has it been realized that there existed animals having the physical characteristics of man in geological deposits hundreds of thousands of years old and even more recent is the discovery that these organisms show physical differences which disappear as the finds approach historic time. Still more subtle is the influence of the thought that man is a late comer and that he is physically insignificant compared with many of the past and present denizens of the earth. The true perspective is humbling. Can the Earth and the Universe still be thought of as especially prepared for the abode of man as taught by the Ancients? Is this view anything more than a childish fantasy? Such problems are beyond the realm of geology, but any one accepting geological conclusions must either give up this belief in its childlike simplicity or remodel it into a problem of stupendous responsibility.

Fortunately out of geology comes a principle of encouragement. Small increments of change will produce tremendous results. A few flakes of snow-fall more each year than the summer's meltings would ultimately give a Glacial Epoch, a slow accumulation of stresses will ultimately produce an earthquake, or the sinking of the continents a foot a century would finally submerge them beneath the sea. Such have been the steps in geologic history.

Geological investigations do not lead to irreligion. Many, if not most, of the leading geologists have been truly religious, for geological phenomena give an insight into the workings of Nature on a scale so large that it sobers the investigator and makes him marvel and resolve to devote himself to the truth as it may be revealed. Geological studies like all other scientific studies do, however, demand a close scrutiny of statements and the discarding of that which is false or misleading, and this has frequently led to criticisms of statements of religious truth in pictorial form based upon incorrect conceptions of the earth, its form, its contents, and its origin.

CHAPTER VI

THE NATURE OF LIFE

By C. JUDSON HERRICK [1]

THE VITAL FUNCTIONS

ONE would like to introduce this theme with a clear and succinct definition of life, but this unfortunately we cannot do either at the beginning or at the end of the chapter. We all know at first hand what it means to be alive, and a critical examination of other kinds of beings that are alive and those that are not reveals some differences which will at least serve as a point of departure for our inquiry.

The living body is clearly a machine. It may be much more than that; but let us first see how as mechanism it differs from other machines. Living substance contains no elementary materials not found in the inorganic world, but in protoplasm these materials are combined in ways not known elsewhere. Something similar is true of every mechanism. An iceberg contains no elements not found in snow or a fleecy cloud, but the berg is not a snowdrift or a cumulus. The elements of which an airplane is built are widely scattered in mineral veins and muddy roads, but these things must be put together in very definite ways to make a flying machine. And this is not all. When the material is assembled according to specifications, it will not pass inspection as an airplane unless it will fly. An airplane, accordingly, must have a certain structural pattern

---

[1] Professor of Neurology, University of Chicago.

and also a definite pattern of action or behavior; and the dynamic pattern is as essential as the structural plan, for indeed these are two aspects of the same thing.

Now, the properties of all natural things are commonly regarded as pertaining to, or at any rate always associated with the material of which they are composed; but we do not know why the element hydrogen has its particular weight and chemical affinities, or why oxygen has a different weight and chemical affinities, or why when these elements are combined in the proportion of two to one they produce water with properties so different from those of either constituent, or why water congeals at 0° Centigrade instead of at some other temperature, or why when these two elements are combined with nitrogen (the most inert element) and carbon in certain proportions they form a dead protein like egg-white, or why when various proteins and other things are united in particular ways the combination is alive. The evidence is that all of these things are true, and the fact that certain combinations of proteins with other substances are alive is no more incomprehensible than is the fact that the properties of water arise from a particular combination of hydrogen and oxygen. The ultimate nature of everything in our world remains mysterious and is not revealed by science, for science knows no ultimates.

Life as the biologist views it is one of the properties of protoplasm just as truly as are its viscosity and specific gravity and there is no more scientific reason for "explaining" it by invoking a metaphysical vital force than there is for postulating a proteid force to account for egg-white or an aqueous force to account for water. The properties of these substances we accept as we find them. Our first concern is to learn what they are, and as our knowledge enlarges more and more of these apparently unrelated properties are seen to be bound up with each other in orderly or lawful fashion.

This, then, is the only kind of "vitalism" that can endure the scrutiny of current biological analysis.

Living substance differs in composition and properties from dead substance. The properties which pertain to life we call the functions of protoplasm. The elementary materials and, so far as we know, the elementary properties (chemical affinities, surface tensions, osmosis, ionization, electrolysis, catalysis, etc.) are all found in inorganic nature, and the act of living is the combination of these elements in distinctive patterns different from those of any dead mechanisms. How living substance was first formed is as yet an unsolved riddle. Dead substances are being continually incorporated into living protoplasm; and this, so far as is now known, is accomplished only through the instrumentality of other living matter, though the possibility that so-called spontaneous generation may be going on to-day cannot be denied.

From this point of view the most distinctive properties of protoplasm, that is, the most essential things about living, are two: first, the machine does not wear out, it is self-repairing; second, it increases in size and complexity of organization. The first is a conservative factor, the second is a progressive factor. There is no mundane dead mechanism that perpetually exhibits these properties; whether the cosmos as a whole does so is too big a problem for the biologist.

Nevertheless these two vital properties are, after all, relative matters. Crystals, mountain ranges and rivers grow in size, and the river may deepen its channel by erosion, extend its length by delta formation, and diversify its configuration in a thousand ways. But sooner or later these physical features, like man-made machines, are corroded and destroyed by the forces of nature. Sun, moon and stars decay, but the trend of life, so far as we know anything about it, is ever toward greater abundance, variety and richness of the process of living. This is the mode of organic evolution.

The capacity of the living body for self-repair is not un-limited, for old age and death are never far from any of us. But not all of the protoplasm dies, and generation by generation youth is renewed and life is enlarged. This increase is not only in mass of population, but also in wealth of living of the mass as a whole and of the individuals. What we mean by wealth of living must be considered more in detail later.

The Vital Energies

It has often been pointed out that the trend of events in the world as a whole is toward the degradation of effi-cient or "free" forms of energy capable of doing useful work to the "bound" or unavailable form. A hot iron under a kettle of water will heat the water and so may do useful work; but when iron and water have reached the same temperature no more "free" energy is available, though the total amount of energy is just the same as it was at the beginning. This is an application of the second law of thermodynamics or energetics.

Applying this law rigorously to its logical and bitter end leads to the disquieting conclusion that the available energy of the universe is gradually running down, so that at the finish all will be as cold and dead as the moon and not so beautiful to look at, were any eye left to see. But here the biologist, who is by natural right an optimist, steps in and says that, within the biological realm and within the range of time spanned by life on our planet (which in all prob-ability has some millions of years yet to run) the matter faces just the other way about.

It is everywhere recognized that there are local and transient reversals of the general process of degradation of "free" energy to the "bound" condition, as when the heat of the sun lifts water from the sea to the clouds, whence it can return this energy as falling rain and water power of streams—very efficient forms of "free" energy. Now, organisms are constantly doing just the same sort of thing and like the clouds they derive most of the energy used in the process (directly or indirectly) from the sunlight, acting in this case on the green leaves of plants. Inert inorganic materials are thus built up into organic compounds like sugar and starch endowed with high latent or potential energy, or they are incorporated into still more efficient living tissues.

There are good evidences that this process of reversal of the cosmic trend toward final degradation of all "free" energy has been progressively increasing during the entire period of organic evolution, for the animal and plant kingdoms to-day are in the aggregate far more efficient in subjugating the forces of inorganic nature and redirecting them in useful ways than were their ancestors whose history is preserved for us in the paleozoic fossils laid down perhaps thirty million years ago.

Man is more competent in this enterprise than any of his predecessors or contemporaries, for with the aid of tools, windmills, dams, engines and dynamos the "bound" energies which he can thus release and step up to a higher plane and so set to work for his own purposes seem boundless. As the electric transformer steps up low potential energy to higher voltage, thus making it (or some of it) available to do kinds of work otherwise impossible, so the human organism is constantly using the low level energies of inorganic and organic nature to run that complex machine that we call civilization. And the progress of the past hundred years in this respect has perhaps been greater than that of the entire previous period of human history. What conquests still lie before? The answer to this question lies quite out of reach of the imagination.

There are two quite different measuring rules which have been employed in gauging the vital processes. Both are quantitative, in a way, as all measurements must be. One measures the amount of energy employed, much as we rate an engine in terms of horse-power, and energy equations of this sort can be written for human and other bodies with a high degree of accuracy—a simple rule of energetics. The other rule measures in terms of pattern of material and process—what is done with the energy. Running round in a circle may be a useful activity, as in an old-fashioned threshing floor; but if one is lost in a fog and wants to get out of the woods it does not get him anywhere.

Practically, then, organisms are ranked as high or low, not in terms of quantity of energy used, but in terms of success in doing useful things with this energy—finding food, mates and other desirables, avoiding enemies, and improving the conditions of living. The energy turn-over of a large tree in a single sunny day is enormous, but we reckon the grub that lives within its bark as a higher organism, for it carries on more varied industries and taps the store of environing energies in more diversified ways. The mere possession of the power of locomotion is of tremendous biological significance. The freely moving animal can go out to meet its world or withdraw from it, and this capacity lies at the root of all the higher functions of self-preservation, self-aggrandizement and self-realization.

CONSERVATIVE FACTORS

The conservative vital forces to which reference has already been made are those which maintain the individuality of the organism and insure its propagation in like form. There is a constant stream of material coming into the body as food and going out as waste, and there is a similar flux of energy, careful measurements showing that the sum of all the energies taken in balances the outgo of energy in work done and the internal processes of maintaining the tissues in working order as precisely as in any other going mechanism.

During these processes the human body takes in, say, two pounds of dry food (plus water and oxygen) daily, but its weight may remain unchanged day by day. The pattern persists. Even during growth, while new tissue is rapidly being added, the pattern is not lost; the personality of the child grows into that of the man without loss of its identity.

Heredity is another conservative factor. The child is himself, not father, mother, or a mean between the two; yet the essential human attributes are passed on and some of the individual and family traits of the parents are always in evidence. The pattern in this case also persists; and in both cases it is a dynamic pattern, not a preformed substance, that endures.

Even so simple a thing as a soap bubble has a transient individuality which endures for a time. If deformed, within certain limits, it will as soon as free to do so restore the original pattern of an iridescent sphere. So a human life is an energetic pattern which expresses itself in bodily features and personality which in a normal environment run true to type so far as free from external restraint and drive in other directions.

The tendency of the organism, if the normal form or pattern of behavior is disturbed, to make such readjustment as will permit the vital processes to continue is known as biological regulation. To some biologists this apparently purposeful readjustment seems so mysterious as to make it necessary to call in the aid of a metaphysical agent or principle, some *élan vital* or entelechy; but to give a thing a highbrow name is no explanation, and in this case the only result is profound narcosis of the scientific spirit and method.

As a matter of fact, regulation of the sort under consideration is seen throughout both inorganic and organic nature. A river is an expression of a certain volume of water flowing over a certain configuration of land under the influence of gravity. A change in the amount of precipitation within its basin or in the topography is immediately followed by regulatory changes in the river itself. So the organism is constantly readjusting to changes in its environment. Thus it expresses its inner nature.

Regulation is merely another name for behavior; for, as Jennings points out, behavior is regulation. Regulation of form has been reduced by Child to similar principles.

The steady trend of all living things toward conservation and repetition of the type is beset by forces from the outside turning it, now this way, now that. Life is interaction. The body must adjust itself to its surroundings. If the environment were uniform, this adjustment would sooner or later be made once for all, the forces of life would come to equilibrium with those outside, and life would cease. Static equilibrium is death. Nirvana is not perfect life, it is the negation of life.

But the environment does change, and to keep alive one must replenish exhausted food supplies and meet the buffetings of a harsh and unfriendly world. The dead machine sooner or later succumbs to the disintegrating forces and is scrapped. The organism, on the other hand, wrests from nature the materials and energies that it needs, keeps them as long as they serve its own purposes, and then returns them stamped with the impress of its own nature. And it continues to command these natural servants to do its bidding and attend upon its requirements as long as it remains alive. This is life. And in proportion as the living body controls the forces of its environment in this way and controls its own behavior in relation to them its life is abundant and efficient. This is our measure of life and the worth of life.

PROGRESSIVE FACTORS

So far attention has been directed primarily to the organism as a machine, deriving its substance and energy from the world about it and by virtue of its internal organization making use of these things for its own good. The most essential differences between this machine and other machines are, first, that it runs but it does not run down or wear out; and, second, that it does not run in closed circles but in ever widening courses in directions that are on the whole progressive from the standpoint of fulness and diversity of life and effective control of its surroundings and of its own destiny. The next inquiry is naturally into the mechanisms actually employed in maintaining this organization and in particular into the apparatus and significance of the progressive movement.

First it should be mentioned that the existence of the progressive trend is by some biologists called in question and the whole organic realm is thought of as drifting aimlessly about as the winds of chance shift hither and yon. But the facts speak otherwise. We have a tolerably good, though by no means complete, record in the rocks of the actual history of life on this planet, and this record tells in unambiguous terms of progressive change from generalized and lowly organized forms to numerous very much higher types of plant and animal life, when ranked according to the criteria already laid down, that is, according to their ranges of capacity for diversified living and getting on in the world. The biologist who is not intellectually myopic evaluates evolution in terms of epochs of time in comparison with which the period of recorded human history is but a single turn of the wheel of destiny. But even within this shorter period human progress in the sense defined is unmistakable to one whose vision is not dimmed by personal, local or transient adversity.

The first progressive factor of living adds to self-preservation, self-development, that is, growth and differentiation. A second factor adds to the uniformities of heredity the variations arising anew in each generation, the survival of successful new forms, and adjustments to new situations, that is, evolution. These are matters, not primarily of quantity of living, but of patterns of living, things that make living more worth while.

THE BIOLOGICAL INDIVIDUAL

The maintenance of the individuality, or in its higher aspects the preservation of the personality, is the primary biological problem. We do not know as fully as would be desirable just what this individuality is, but we do know some of the instrumentalities employed to preserve and enlarge it.

Living has been described as *the process of continuous adjustment of the internal activities of the body to the forces of surrounding nature.* Since the vital energies all come directly or indirectly from the outside, it is evident that the immediate effect of external agents upon protoplasm is a matter of critical importance. This has been much studied and manifestly much of living is direct response to stimulation.

Protoplasm, in common with nitroglycerin, photographic plates and many other things, is irritable. The immediate reaction of the protoplasm to an irritant is known as excitation, which is a rather complicated upset of chemical and physical equilibrium with pronounced change in electric state and release of energy. These changes are more or less readily transmitted from the point stimulated through the surrounding protoplasm so that parts of the body, perhaps in very remote places, may be excited to activity by the initial stimulus. The course of the excitation during this transmission can readily be followed by reading on a galvanometer the related electrical change, known as the action current or negative variation. Other and still more delicate methods are also available for this purpose. Many of the tissues of the body, such as nerve cells and muscle fibers, are storehouses of latent energy in unstable equilibrium and set on a trigger so that their reserves of energy may easily be released. The energy of protoplasmic transmission may pull these triggers and discharge the reserves, thus activating now one, now another, of the organs of response, just as the movements of all the trains of a big railroad system are controlled from the dispatcher's office.

Now, ordinary protoplasm is not a very good conductor of these excitations and the efficiency of the process runs down progressively as the transmission emanates from the excited point. Thus arise physiological gradients running from high activity at the place of initial excitation to extinction at more or less remote points. The high points of these gradients have been shown to exert more or less of regulatory control over the activities of all other parts of the body within the range of their influence, that is, they are physiologically dominant over these other parts. Parts of the body which are highly excitable, such as the head ends of freely moving animals, are thus centers of dominance with reference to all other parts, so that they control the movements of the body as a whole. There may be many subsidiary centers of dominance distributed throughout the body, which exert a control of more restricted range, and so the whole body comes to be knit together as a system of excitation-conduction gradients whose pattern in its main features is an expression of the arrangement of the sense organs and organs of response, but whose details will vary from moment to moment with changes in external stimulation and internal processes.

These physiological gradients, or gradients in rate of living, are what hold the body together as a working unity and determine its individuality and the pattern of its form and behavior. There are other important factors in this process, such as the mechanical arrangement of parts, the transportation of nutrient and other materials through the blood stream, and wide-spread chemical effects of the products of the glands of internal secretion; but the excitation-conduction gradients are the most important and probably the primary factors in the unification of bodily activities.

This conception, here very briefly and inadequately outlined, has been elaborated by Professor Child on the basis of an extensive series of experimental studies of the gradients in lower organisms. In higher animals, where there is a well-developed nervous system, the same principles apply, but in modified form, for here the nerves are excellent conductors and the range of action of the centers of physiological dominance is greatly enlarged. The centers of highest dominance, moreover, are transferred from the sensory surfaces at the periphery to the adjusting centers within the brain, and in the case of man to the cerebral cortex in particular.

The physiological gradients and the organs of neuromuscular control thus comprise the chief (though by no means the only) apparatus of unification of the bodily activities and preservation of the individuality. And this control is not that of a *deus ex machina* arbitrarily imposing his decrees upon obediently responsive tissues, but it is a control determined from within by the action and interaction of part upon part in definite patterns of systems of excitation and transmission, in coöperation with the transportation from part to part through the body fluids of foods, wastes and powerful internal secretions and various mechanical factors of bodily organization.

In plants and lowly animals the dominant centers of the physiological gradients are regions of greatest physiological activity, that is, they are more alive than other parts; and their regulatory control is exerted through the instrumentality merely of their higher rate of living. In higher animals with well developed central nervous systems the matter is much more complex, for the regions of highest dominance here are the correlation centers of the brain, and particularly the cerebral cortex. The nervous mechanisms involved here are sufficiently well understood to justify the statement that probably the dominance of the cerebral cortex over the rest of the human body is maintained by an elaboration and modification of those physiological principles of excitation-conduction gradients observed in plants and jellyfish. But this theme is too difficult and involved to be developed here.

MIND AS A BIOLOGICAL FUNCTION

Now, some of the functions of the cerebral cortex have the unique property that they are aware while they act; there is conscious experience of some sort during the progress of the transmission of the physiological excitations through the inconceivably intricate mazes of cortical associational pathways. This is the greatest riddle of life, for the exact relation between this awareness and the associated physiological function opens up problems which far transcend the limits of biology.

But looking at the matter from a strictly biological point of view and discarding so far as practicable philosophic or other preconceptions, thinking, feeling and willing as we know them in common and scientific experience are functions of the living brain (or of parts of it) in just the same sense that contraction is a function of muscle. This is no more an objectionable materialism than is the colligation of any function with its organ.[2] The ultimate ex-

---

[2] A discussion of the philosophical problems associated with the mind-and-body question would fall quite outside the scope of this

planation of every vital activity lies beyond our reach, and in all of these cases the biologist can only say, "I do not know."

The adoption of this obvious and naïve (though pragmatic, not ultimate) conception of the relation of body and mind at once opens up fields of inquiry by the biological method which hitherto have been posted, "No thoroughfare," against invasion from this side. Defining mind in the traditional sense as awareness of some sort,

---

essay, but fear lest the condensed and dogmatic form of statement may lead to misunderstanding prompts the addition of a few words parenthetically.

The reader is requested to bear in mind that this chapter begins and ends in the biological field, and the writer believes that within this field his argument can be adequately supported by citation of data of observation and experiment. Man as a social organism can be (though rarely has been) evaluated by biological criteria as truly as a hive of bees or a colony of polyps. And when so evaluated the subjective and the objective aspects of his life knit together with each other and with his organic and inorganic environment without sacrifice of either aspect or subjugation of one to the other.

In a strictly biological survey of human life, to treat mind as an epiphenomenon or as a parallel phenomenon which can be left out of the reckoning in the study of cause-and-effect sequences is a scientific travesty. The most familiar facts cry aloud in refutation, for mind is the most obvious and the most potent biological factor in the realm of human nature.

But to say that the mind of our experience is a function (or property) of the living body—for which the biological evidence is in my opinion adequate—is far from defining what mind and matter really are and what may be the ultimate relation between matter and its properties, between a living body and any of its functions. Here is a philosophical problem of the first magnitude which probably must wait upon the accumulation of further experimental data. These fundamental questions the biologist—as biologist—is inclined to lay aside as beyond the range of the present technique of his specialty. Living bodies as we know them and their functions (all of them) form the material of his science. These are natural bodies. Whether there may be spiritual bodies with other properties not open to investigation by the scientific method, is again a question which—as biologist—he cannot discuss.

the mental functions are related with other bodily processes in causal sequences. The evidence for this is positive and unambiguous. Bodily processes, intoxication, etc., affect the mind and mental acts control bodily acts. Mental labor commands a higher wage than manual labor and the energies of mind are the most potent of all biological factors in giving mankind his control over the rest of nature.

Mind as cause is the most efficient of all of the progressive factors in evolution, and the transfer, within the span of the (relatively) few thousand years that mankind has occupied the earth, of the field of most significant evolutionary advance from the lower biological to the higher biological (that is, mental) plane has sped up the progressive movement, as previously defined, to an unprecedented degree. Especially during the five thousand years (more or less) during which mankind has been developing a scientific knowledge of nature and the apparatus of social control and social heredity, progress in mental culture has largely replaced physical prowess as an evolutionary factor.

The key to this progressive evolutionary movement, as of all others, is *control*, but here magnified and glorified through the instrumentality of a cortical mechanism in which the dominant element of behavior is not reaction, but action, not immediate response to environmental impacts, but generalizations of experience, deliberative behavior, forecast of future events, fabrication of ideas and ideals and the shaping of conduct in conformity therewith.

Thus control of nature, control of social organizations, and control of self arise naturally within an evolutionary process whose end we cannot yet see. But the trend is apparently toward higher individual efficiency with corresponding richer life, and an enlargement of the individual interests to encompass a wider social group with realization that the highest individual welfare is bound up with the subordination of certain personal, family, and (let us hope, in the end) national interests to the larger social good.

"Ape-and-tiger" methods of evolution are already out of date in human society, though unfortunately many individuals and many communities have not yet found this out. The personal ideal and the social ideal have grown up within an evolutionary process, and these are the chief instruments of further evolutionary movements. The direction to be taken by these movements will be determined very largely by the nature of these ideals. Fortunately this is a matter which can be controlled by education and the other instruments of social progress far more rapidly than any of the more archaic modes of physiological evolution can operate.

CONCLUSION

We have seen that physiological control and the resulting biological efficiency, from the lowest organic levels to the highest, are brought about through mutual action of part upon part and mutual subordination of part to part. Regions which are most active, most alive, naturally exert the stronger action and so come to manifest more or less dominance or control over the activities of other parts. This is true throughout the physiological realm.

If mind is regarded as a bodily function knit in with all the others on the biological level, then the mental functions take their places in the natural physiological order, and in mankind they dominate the other bodily functions and in turn are influenced by them. Personal efficiency, under the control of personal ideals and social obligations and sanctions, appears to be the criterion of further human evolution, a criterion which is congruous with, and probably derived from, the elementary physiological factors mentioned in the preceding paragraph.

CHAPTER VII

PLANT LIFE

By JOHN M. COULTER [1]

PLANTS can be regarded from many angles. From the beginning of human history man has found plants useful. In fact, all animal life, including man, is dependent upon the work of plants for its existence. Many have also found plants beautiful, the so-called "vegetation" covering of the earth's surface adding attraction to a landscape. To appreciate plants, however, one must look beyond their usefulness and beauty, and discover how they live and work. This is the scientific attack, and it will be of interest to note briefly its historical development.

The first scientific attack upon plants was the attempt to classify them. The first attempt at classification was arbitrary, as arbitrary as the classification of words in a dictionary by the accident of the first letter. Gradually, however, classification became more and more natural, until at present plants are classified according to their relationships, as indicated by the resemblances and differences. The study of classification, therefore, has organized the plant kingdom into a kind of family tree, according to "blood relationship."

The next attack was upon plant structure. The bodies of plants were described as made up of certain organs. With the invention of the microscope, the attack upon structure became more intimate, and tissues and cells were uncovered. Finally the attack extended beyond the mature

[1] Professor of Botany, University of Chicago.

body to the developing body, and all the stages in development, from egg to mature body, were uncovered.

The next attack was upon plant work. Structure means little if its significance in work is not discovered. One may examine all the parts of a locomotive, but unless he sees the locomotive in action, the structures mean little. Conversely, to see a locomotive in action is not to understand it, unless the structure is examined. In this way, structure and function are bound together. Both must be investigated to understand either.

The next attack was upon the responses of plants to their environment. It was discovered that plants are dependent upon their environment, and respond in a variety of ways. Different plants are adjusted to different environments, and for this reason the vegetation of the earth is exceedingly varied. Why can certain plants live where others cannot? After the investigation of the responses of individual plants to environment had begun, it led to the larger view of what are called plant associations or communities. It was found that certain kinds of plants naturally associate to form a community, and environmental study advanced from the individual to the group. It is really a study of plant sociology, and it has proved to be full of suggestions. Plants may be studied as individuals relating themselves to their surroundings, just as a human individual may be studied as he adjusts himself to the conditions of life in a city; or they may be studied in "vegetation masses," such as forests or prairies, just as groups of people in a city may be studied as they adjust themselves to other groups. One great natural vegetation mass is of such practical importance that it has developed the special subject of forestry.

The next attack was upon the diseases of plants, which so often ravage our crops. The chief causes of these diseases are other plants, so that this study involves a knowledge of the structure of two kinds of plants, those that attack and those that are attacked. It also involves a knowledge of two kinds of work, the work of the plant when in health and its work when diseased.

The most recent attack is known as plant-breeding, and it has become of great scientific and practical importance. It means the growing of plants, generation after generation, under observation and control, and trying to discover the laws of inheritance, which we usually call heredity. This is the great scientific importance of plant breeding. Its practical importance comes from the fact that the scientific work has suggested methods of improving our old plants, producing new ones, and guarding our crops against disease and drought.

These six aspects of plants do not exhaust the list, but they are conspicuous illustrations of the fact that there must be a synthetic attack upon plants if we are to discover the explanation of their life and work. In the following pages no attempt will be made to present the results of these various methods of attack. Two conspicuous results will be selected, which may be said to have a human interest as well as a scientific interest, and are conspicuous illustrations of the life and work of plants, namely, plant evolution and plant sociology.

Plant Evolution

The plant kingdom, for convenience, may be regarded as an apartment building of four stories, each story representing a certain level of advancement in the development of structures to meet life conditions. The conclusion that these four levels represent the evolutionary succession of plants is reached by focussing upon them our knowledge of structure, of fossil records, of the responses to environment, and checking up by experimental work. Changes that have been inferred have often been produced experimentally.

These four levels will be described briefly, thus developing a historical picture of the plant kingdom, corresponding to human history.

The Primitive Plants

The most ancient plants are certainly not known, but we must infer that our simplest plants are the most primitive ones in our present flora. This most primitive existing group is made up of the Algæ. This does not mean that they were necessarily the first plants, for plants that have disappeared, or that we have failed to recognize, may have preceded the Algæ. In our present flora, however, the Algæ appear to be the forms that have given rise to the other groups. Algæ are of little practical importance, but they are of very great scientific importance, because they illustrate the beginnings of the plant kingdom, and show how the important kinds of plant work are provided for in the simplest way.

If Algæ are the primitive plants, it follows that the plant kingdom began in the water, for Algæ grow in water or in very moist places. It seems to be true, also, that the most primitive Algæ, as well as those that gave rise to the higher plants, lived in fresh water. Algæ are commonly called "seaweeds," but those that live in marine conditions are not the most primitive, nor have they given rise to higher plants. So far as we know, therefore, the plant kingdom began in fresh water, and naturally it had to begin its work of food manufacture before animals could exist. To live in water as a medium means that all the structures and habits of such plants must be adjusted to water. That plants living in the water are relatively simple is illustrated by the fact that when plants live in the air they must be protected against drying, and this involves protective structures that water plants do not need.

The chief reason, however, for regarding Algæ as our most primitive group of plants is that the body of the simplest Algæ consists of a single cell. Since a cell is the unit of structure, the body of these plants consists of a single unit. This single cell, however, does all the kinds of work done by the most complex plants. These kinds of work are chiefly food manufacture, nutrition of the one-celled body, and reproduction. In many-celled plants these functions become more or less distributed in certain special organs. In other words, as the plant kingdom progresses, it is not the kind of work that advances, but the machinery for conducting the work.

It is important to recognize the primitive method of reproduction. Any living cell, when exposed to certain conditions, divides, resulting in two cells. When the body consists of a single cell, as in the lowest Algæ, this division results in producing two individuals; that is, reproduction results from cell division. It is an interesting fact that in this case the parent cell merges entirely in the bodies of the two progeny cells. The parent cell ceases to exist with every act of reproduction.

From this start, a one-celled body and reproduction only by cell division, the Algæ have progressed in developing a more complex body and in the evolution of reproduction. They have not progressed very far in body structure, but they have carried the reproductive methods to their final expression. The evolution of methods of reproduction, therefore, was completed by plants while the body remained relatively primitive. Of course the machinery of reproduction becomes more and more intricate beyond the Algæ, but they have worked out all the methods.

The evolution of the body proceeded as follows: When the one-celled body divides, the tendency is for the progeny cells to remain in contact, and as divisions proceed the result is a group of one-celled individuals, known as a colony.

Often these colonies are held together by a kind of mucilage formed from the material of the cell walls. In this way there is an enforced association of one-celled individuals. This living together gradually results in the beginning of the appearance of a "division of labor." The common expression of this is that certain cells act as holdfasts or anchor-cells for the colony, while the others are active in food-manufacture. In other words, the individuals of a colony cease to be entirely independent and become mutually dependent. This gradually merges into an individual of many cells, and this is as far as most of the Algæ have gone in body structure. In general, this many-celled individual is a filamentous body, simple or branching, since free exposure of each cell to the water medium is of advantage. Occasionally the body is a sheet of cells, like the well known "sea-lettuce," common on the seashore. The progress of the body, therefore, can be summed up in the statement that it advances from a single cell to a filament of cells, or occasionally a sheet of cells.

The evolution of reproduction involves many steps, which can be indicated briefly. When the body becomes many-celled, cell-division does not result in reproduction, that is, a new individual, but in growth of the body. For a new individual to arise from such a body, therefore, something more than cell-division is necessary. The living material of a cell, known as protoplasm, is organized into a unit known as the protoplast. Usually in plants the protoplast forms a wall about itself, and it is this wall that induced the old biologists to call the unit structure a cell. When a many-celled Alga encounters hard conditions of living, the protoplast becomes more and more inactive and finally shrinks away from the wall covering. As a result the wall eventually breaks in various ways, and the protoplast escapes. This escape protoplast continues to do the same things as those that have not escaped, notably it divides, and divisions continuing a new individual is formed. There is no difference in the kind of work done, but in the opportunity to have this work result in a new individual. These escaped protoplasts are called "spores," and therefore a spore may be defined as an escaped protoplast whose divisions result in a new individual. The first step in the evolution of reproduction, therefore, is the appearance of spore reproduction, associated with the appearance of many-celled individuals.

The next step in advance is even more significant. Spores are produced when conditions for food manufacture begin to be somewhat unfavorable. These conditions become more and more unfavorable as the growing season declines, until at the very end of the season a new situation develops. The protoplast does not escape promptly, but divides within the cell, and divisions continue until the cell may contain numerous very small protoplasts or spores. When these minute spores escape they seem to be too small to start divisions and form new individuals. Occasionally such a spore makes the attempt, resulting in a very small dwarf, proving that it is a spore. In general, however, they behave in a very striking way. They come together in pairs and fuse, the result being a fusion cell that is vigorous enough to produce a new individual. This fusion process is the sex act, and the fusion cell corresponds to the fertilized egg. The fusing cells are called "gametes," corresponding to what are more commonly known as sperms and eggs. This situation shows, therefore, that gametes have been derived from spores, and this is the origin of sex, or at least the origin of sex cells. The pairing of these sex cells shows that they are differentiated into what we call male and female cells, but since they are alike in appearance and behavior, the sexes cannot be distinguished by any method now available.

The next advance is the differentiation of sex, meaning

that the pairing gametes gradually become unlike in appearance and behavior, so that sex can be distinguished. When protoplasts escape as spores or gametes from the bodies of Algæ, they encounter the water medium, and the usual response is the development of cilia, which are swimming appendages. Spores and gametes, therefore, swim about actively for a time. When gametes are alike, both the pairing gametes are small and actively motile. Gradually, however, one of the pair becomes larger than the other, and its activity correspondingly declines, until finally it becomes relatively large and passive, while the other remains small and active. When this stage is reached, the sexes are distinguishable, and the pairing gametes are known as eggs and sperms. The advantage of this differentiation is obvious. The increase in size of the female gamete (egg) has to do with increased food storage. When the egg is fertilized, therefore, there is a store of food to supply the needs of the embryo. In other words, this differentiation gives to the progeny a better chance of initial development.

When sex-differentiation is first attained, any cell of the parent body may produce eggs or sperms. The next advance is a differentiation of the cells of the parent body into those that work for the body and those that produce gametes. This means that sex-organs are developed, composed of cells that are never a part of the working body.

The final stage reached by the Algæ is the separation of the two kinds of sex organs on different individuals, resulting in male and female individuals. This is the climax of the evolution of reproduction, and as yet plants and animals have gone no further.

The evolution of reproduction by the Algæ may be summarized in the following steps: Starting with reproduction by cell-division alone, it advances to the origin of spores, the origin of sex, the differentiation of sex, the differentiation of sex organs, and finally the differentiation of sexual individuals. This does not mean that each step in advance involves the abandonment of the previous methods, for reproduction by cell-division, by spores, and by gemetes is used even by the highest plants. It is evident that if Algæ have worked out the whole program of reproduction, while retaining simple bodies, the evolution of the higher groups will stress the evolution of the body, including the machinery of reproduction.

A brief statement concerning certain relatives of the Algæ may be suggestive. These relatives are the Fungi, characterized by the fact that they have lost the power of food-manufacture, and have become dependents, just as are all animals. This dependent habit of the Fungi is associated with what may be called retrogressive evolution of reproduction. Starting with well-developed sex-organs, as in the higher Algæ, the Fungi gradually retraced all of the advancing steps of the Algæ, so far as sex is concerned, until in the so-called "higher Fungi" sex has disappeared, at least from any ordinary expression. The Fungi, on the other hand, have certainly stressed spore-reproduction.

The First Land Plants

It was stated that the Algæ live exposed to water as a medium, and that their structures and habits are explained by this fact. To live on the land means exposure to air as a medium, and the structures and habits of land plants are explained by this fact. The danger of exposure to air is the loss of water by the plant. It must lose water to the drying air, and unless there is some check the loss will be greater than the supply, and death will ensue. When an Alga is removed from water and exposed to air, it dries out quickly and perishes, for there is no check to the very rapid loss of water from the protoplasts. If water plants are to become land plants, therefore, they must acquire the air habit by developing protective structures. This is just what certain Algæ did, and in doing so they became so different in structure that they ceased to be Algæ.

This may be spoken of as the conquest of the land surface, and the pioneers in this conquest are the Liverworts, whose relatives, the Mosses, are much better known to most people. The Liverworts are not conspicuous, and therefore not noticed by most people, but it is believed to be the group that acquired the land habit and that has given rise to all the higher plants. They may be called the amphibians of the plant kingdom, for they connect the water forms with the land forms.

Remembering that to acquire the land habit, with its danger of air-exposure, involves the reduction of this exposure to a minimum, it is easy to understand the success of the Liverworts. In the first place, the start was made with Algæ whose bodies were sheets of cells rather than filaments. Such a body is compact, with only two sides of a cell exposed, and if the sheet of cells is more than one layer of cells thick, as is usual among Liverworts, the outside cells have one side exposed, and the cells within are not exposed at all.

In the next place, the body lies flat, so that only the upper surface is exposed freely to the air. This position, added to the fact that the body is usually lying upon a moist surface, results in the least possible amount of danger from exposure to the drying air. When the Liverwort body is several layers of cells thick, the outermost layer is modified and becomes the protective layer known in all plants and animals as the epidermis. This epidermis is in effect a waterproof layer, not to prevent water from entering the plant, but to prevent water from leaving the plant. A compact, prostrate body, ensheathed by an epidermis, is well equipped for exposure to the air, especially if the air is not very dry. Not only is the working body jacketed by an epidermis, but the sex-organs are also jacketed for the same reason, which makes the sex-organs of this group much more complex structures than the sex-organs of the ordinary Algæ.

One may picture how a gradually increasing exposure to air might have occurred, beginning with occasional exposures on muddy flats, and, by gradual shoreward migration, ending in continual exposure. Even when exposure to air became continual, it must have been for a long time in conditions of shade and moisture, for life "in the open" means extreme danger from loss of water.

The Mosses are much more abundant and conspicuous than the Liverworts, and were derived from them, but they specialized so extremely in a certain direction that they blocked their own progress. It was the Liverworts that arose from the Algæ by acquiring the land habit, and gave rise to the next higher group of plants, which occupy what we have called the third story of our apartment building.

The First Vascular Plants

After the land habit was acquired, the original land groups (Liverworts and Mosses) remained lowly plants, chiefly in comparatively moist and shaded localities. Their great contribution was in their developing the ability to live in the air at all. The next advance was for plants to "rise from the ground" and learn to "live in the open." This problem was solved by the third great group of plants, comprising the Ferns and their allies. Just as there is a gradual transition from the Algæ to the Liverworts, there is also evident a gradual transition from the Liverworts to the Fern group. The contribution of the latter group to the evolution of plants was the development of a vascular system. Initial stages of such a system are very evident in certain Liverworts, but when the system became distinctly developed, the plants became members of the next higher group.

Vascular tissue means a tissue composed of vessels. These so-called vessels are equipped to conduct water, in addition to other work. Of course water is conducted through the bodies of Liverworts, but the vascular tissue conducts it with more rapidity and precision than any other tissue. The difference between water-conduction in a liverwort and in a plant with vascular tissue may be likened to the difference between water working its way through a swamp and water moving in definite channels. It is this vascular tissue which enabled plants to rise from the ground and attain to the height of trees. The exposure to the drying air and the distance from the water supply in the soil are both greater, but the rapidity of water movement meets these situations. The vascular tissue is commonly called "wood," so that the group we are considering might be called "the first woody plants."

Associated with the vascular system are two important organs, without which the vascular system would be meaningless. One is the root, which is organized for water intake. There would be no use for a water conducting system unless there was a water intake. This group, therefore, represents not only the first vascular plants, but also the first plants with roots. The other organ associated with the vascular system is the foliage leaf. It is these leaves that do the essential work of food manufacture, and it is this work that demands a water supply, and this supply is distributed to the working cells of the leaf by the so-called leaf veins. The vascular system, the roots, and the leaves must all be considered as parts of one system. The root represents the water intake, the vascular system of the stem represents the water conduction, and the leaves represent the water distribution.

Incidentally it may be of interest to know that the vascular or woody tissue, being the most enduring kind of tissue in the plant body, has enabled us to recognize and classify vascular plants as fossils, and we find that their history extends as far back as our fossil records have reached.

The Seed Plants

The plant kingdom culminates in the group of plants characterized by producing seeds. This group is the most conspicuous one to-day, for it makes up nearly all the vegetation that one ordinarily sees. The vegetation covering of a landscape is practically all composed of seed-plants. There is no occasion to define a seed, for it is a very complex structure. Its significance lies in the fact that the seed protects the young embryo plant in a way that no other group has secured. Within the hard seed coat the embryo lies dormant until suitable conditions awaken it to activity, and it escapes from the seed to establish itself as an independent and growing plant. It is an interesting series of events that puts a growing embryo into the dormant stage, to be awakened by another series of events.

There are two main groups of seed-plants, one represented by the pines (the so-called "evergreens"), and the other including the plants that produce flowers. The pine group is the ancient seed-plant group, while the flower group is relatively modern, and is far the largest group of plants to-day. They are the most advanced, the most recent, the most conspicuous, and the most useful of plants. The vegetation that covers the earth is in the main flowering plants, and when to this is added the fact that they are almost the only plants that men use, it is not strange that they were once thought to be the only plants worth studying. It was for this reason that Botany was once thought to be merely a study of flowers, for the flowers are used in classifying this great group.

The significance of the flower lies in the fact that it is a mechanism for securing pollination by insects. The old method for pollen transfer was what is called wind-pollination. This is a very wasteful method, for it calls for showers of pollen to be sure that a few pollen grains may land in the right spots. Insect-pollination secures a definite transfer of pollen from where it is produced to where it can function. It is this adaptation to many kinds of insects that has resulted in the remarkably diversified development of flowers.

Summary

This very cursory glimpse of the evolution of plants must be understood to represent only certain outstanding features that can be appreciated by those who have only a general acquaintance with plants. The same conclusions are confirmed by a study of the more intimate structures. It should be remembered also that the outline presented is not merely a result of the studies of structure, but also studies of the fossil records and of plant behavior, all tested by very much experimental work.

Plants began as one-celled individuals living in fresh water as a medium. In that situation they evolved the whole program of reproduction, but in the medium in which they lived the structure of the body is known to have remained simple.

The gradual emergence of plants from the water and their exposure to air as a medium resulted in a much more complex body structure to guard against the danger of drying out.

After the land surface had been occupied by plants with prostrate bodies, the evolution of the vascular system made it possible for plants to rise from the ground and stand upright. This resulted in greater water loss, but this was compensated for by the rapidity of water conduction and by the better exposure to the air which supplied essentials for food manufacture and respiration.

The final stage was reached in the better protection of progeny with the evolution of seeds, and in the more certain production of progeny in the evolution of flowers in connection with insect pollination.

PLANT SOCIOLOGY

Plants may be studied as individuals engaged in various kinds of work, or they may be studied as groups associated in villages and cities. It is true that the earth is covered by individual plants, but it is also true that these plants are associated together in various ways, forming what may be called plant communities. It is this community life of plants that will be considered here briefly.

Plants are not scattered indiscriminately over the surface of the earth, regardless of one another and of the conditions for growth. It is recognized, for example, that there are forests, prairies, and swamps, each of which represents an association of plants that characterizes it. Into each association certain plants are admitted, and from each association many plants are excluded. Any set of conditions for plants is said to be a "habitat," a place that certain plants inhabit. Each habitat, therefore, has its own association of plants. There is no need in this connection to define the combination of factors that determines each kind of habitat, such as available water, temperature, soil, etc.

When a plant association is visited, it may be looked upon as a community whose population consists of plants. There are certain general features in the community life of such a population that become evident at once. One of the most obvious facts is that certain individuals dominate and give tone to the community. For example, a forest association is dominated by the trees, and often by one or two kinds of trees. This is so evident that many people think of a forest as consisting of trees only, when in fact they are only part of a large population. In the same way, a meadow is dominated by grasses, so that to many it seems to be almost exclusively a grass population. Thus each association usually has its dominating individuals that characterize it. This fact has a very interesting corollary. The rest of the plant population must adjust itself to the dominant individuals. For example, in the forest population the other plants must adjust themselves to the dominating trees. Very many of them are so constituted that they can live in the shade of trees; while others, like "spring flowers," by means of underground storage of food in roots or stems (tubers, bulbs, etc.), can spring up rapidly and come into flower in the short period between the first warm days of spring and the full foliage of the trees, thus finishing their work before the shade becomes dense.

Another notable feature of a plant community is that the nearest relatives are the keenest competitors. If a certain kind of plant has established itself in a community, it is very difficult for a nearly related plant to obtain a foothold. It must not be thought that the competition referred to, whatever it may be, is of the active sort, but the word at least figuratively describes a situation. This fact contains some very practical suggestions. Our worst "weeds" are not members of our native population, but immigrants. In various ways, the native plants of foreign countries become introduced into America. If they find near relatives in our native population, they are not heard of as weeds; but if they find no near relatives, they are probably freer from competition than they were in their native country, and may become a pest. The important suggestion, however, is that the more kinds of plants there are on a given area, the larger will be the total plant population. In cultivating only one kind of plant at a time on an area, therefore, we are reducing the possible plant population to its minimum.

The most important fact in reference to a plant association is that it is not permanent on a given area. In general when a plant association lives for a time upon an area, that area becomes increasingly unfavorable to it, until gradually it is succeeded by another plant association. Almost any plant association finally makes conditions unfit for itself, and at the same time more fit for some other association. This succession of plant associations may be illustrated by the succession of human communities. Pioneer conditions bring together a characteristic association of individuals, but the conditions do not remain pioneer, and become favorable for another association of individuals, and this kind of succession may go on, until the series of associations can be traced from the pioneer association to the metropolitan association. This means that each plant association can reveal the succession of associations that preceded it, and also the succession of associations that will succeed it. In other words, the most important thing about a plant association is the history and prophecy it contains.

It is evident that there may be many kinds of succession, dependent upon the habitat. The start may be on bare rock, on sand, on clay, in a drained swamp, or in an undrained swamp, and then each kind of succession will follow. It is also evident that succession cannot go on indefinitely, but that some final association will be reached which is called the climax association for that region. In general, some type of forest is the climax association, but there are obvious reasons why that type cannot be reached in certain regions.

Since forests represent the most important natural vegetation, an illustration of forest succession will be given.

It will serve to illustrate not only an important succession, but also the facts that must be considered in any effective study of forestry. One of the best known forest regions is the white pine region of Northern Michigan, from which the trees have been swept, with no thought of their continuance, and the evolution of this forest will indicate the forest problems in general.

The succession of plant associations which led up to the white pine forests started on sand, rock, clay, or in swamps, but the series beginning on a sandy beach will be used in the illustration. The first stage was the lower beach, washed by the summer waves, and therefore with no vegetation, but with an accumulation of sandy soil. The second stage was the middle beach, rising higher above the water, and therefore washed only by the larger winter waves. This freedom from waves during the summer permitted growth of certain annual plants, whose bodies added some humus to the soil. The third stage was the fossil beach, that is, a beach that was once washed by the waves, but is now beyond their reach. This continual freedom from wave action permitted the growth of more plants, and therefore resulted in the accumulation of more humus, but the soil would still have looked rather bare, as the plants would not cover the surface. The fourth stage was made possible by the accumulation of humus, and it is called the heath stage, for plants of the heath family and their associates covered the ground. At this stage, for the first time, the plants covered the ground so thickly that competition among individuals began.

With the further accumulation of humus, the fifth stage became possible, that is, the pine forest stage. Gradually the pines invaded the heath, first the jack pine, then the red pine, and finally the white pine. It was at this stage that men caught the succession and destroyed the pines. When a forest fire swept through the pine forest, it not only checked the succession, but often set it back. If the fire was prolonged and intense, it not only destroyed trees, but also much of the humus, and in such a case the succession might be set back to the heath stage. This would mean a long accumulation of humus before the pine forest could come in again.

The important fact, however, is that the white pine is not the climax forest for that region, for it has the curious habit of what may be called race suicide. Its seeds do not germinate well and its seedlings do not thrive in the shade, so that when a deeply shaded white pine forest is established, it cannot perpetuate itself. But this shade is favorable to the seeds and seedlings of the maple and beech, and therefore these trees gradually supplant the white pine, and the maple-beech forest (a hard-wood forest) is the climax association for that region. It follows that the problem of forestry in the white pine region is not merely a fight against the devastations of men, but more fundamentally a fight against the race suicide of the white pines and against the encroachment of the hard-wood trees.

This is an illustration of but a single forest succession out of a great many. For example, in Oregon and Washington, where the conifer forests are so conspicuous, they are the climax type, and there is no danger of an invasion by hardwood trees. The conifers of that region do not commit race suicide, and the hardwoods are not favored by the winter rains and dry summers.

Another interesting feature of plant communities is what may be called their contests with one another. These contests are friendly and consist essentially in getting control of what might be called the "market." Moreover, they are not conducted by communities as a whole, but exclusively by the dominant plants, whose camp followers come in whenever any new vantage ground has been gained. In a general way contiguous communities develop a state of equilibrium, visibly represented by what may be called neutral ground, and this may be respected for a very long time. If any new condition, however, results in weakening the powers of resistance of one of these communities, the other will be sure to encroach. A striking example may be obtained from what is going on now in Texas. As is well known, that state contains vast grazing areas, which means very conspicuous grass communities. These communities naturally are so vigorous that they have kept back the shrubby vegetation to a large extent. Years of grazing, however, have diminished the strength of resistance of the grass land, and to-day the characteristic shrubby vegetation of the state, such as mesquite and its allies, is slowly but surely encroaching upon the grassland. In this case man has entered upon the situation, and has thrown the balance of power in favor of the undesirable vegetation.

The illustrations of plant life that have been given are simply representative of innumerable activities. The important thing to recognize is that plants are alive and at work. Many seem to think that animals are alive and plants are not, their test of life being locomotion. This is far from true, for some plants have the power of locomotion while some animals do not. Both plants and animals are living forms, and the laws of living that animals obey must be obeyed also by plants. The recognition that plants are alive and at work gives meaning to their forms and structures and positions. For example, the form and structure and position of a leaf have no meaning until it is discovered how these things enable the leaf to do its work. Although many different kinds of work are being carried on by plants, all the work may be put under two heads, nutrition and reproduction. This means that every plant cares for two things, namely, the support of its own body and the production of other plants like itself. In the cultivation of plants nothing is so important as to know about their nutrition and reproduction. Knowledge of the nutrition of plants enables one to secure vigorous plant bodies, and knowledge of the reproduction of plants enables one to secure desirable races of plants. In fact, the material welfare of the human race is largely based upon the life and work of plants.

CHAPTER VIII

ANIMAL EVOLUTION

By HORATIO HACKETT NEWMAN [1]

THE STATUS OF EVOLUTION IN AMERICA TO-DAY

NOT since the days immediately following the publication of Darwin's *Origin of Species* has popular interest in the Principle of Organic Evolution run so high as it does to-day. The reason for this is not far to seek. The opponents of the principle have advertised it in a highly effective manner through their sensational modes of attack. The newspapers and other popular prints have been loaded with anti-evolutionary propaganda, so bitter in invective and withal so entertaining that it has usually been accorded front page space. In many instances, however, where it was not possible to obtain sufficient space in any other way, whole pages of propaganda were published and paid for at advertising rates in some of the most widely circulated of our newspapers. The result has been that vast numbers of people, who perhaps had never heard of evolution or at least had never before bothered their heads about the matter, had the issue repeatedly forced upon their attention.

Even more effective in exciting popular interest than newspaper propaganda has been the attempt, more or less successful in a few instances, to use the legislative machinery of several of our less progressive states as a weapon of attack. Various measures whose aim has been to render

---

[1] Professor of Zoölogy, University of Chicago.

the teaching of the theory of evolution a statutory offense have been fought out in the various legislative bodies amid scenes whose entertainment value has probably excited the envy of the leading purveyors of popular amusement. All this has been played up in the metropolitan press with especial emphasis upon the more amusing episodes. The result is that the mention of evolution nowadays appears to be good for a laugh in almost any company. Ridicule, when skillfully used, is one of the most effective of controversial weapons; but to remain effective it must be aimed aright. In this case the weapon seems to have back-fired with considerable violence, for the laugh is usually at instead of with the leaders of the attack upon evolution.

This attack has been carried on at so low an intellectual level that there has been pronounced and doubtless unexpected reaction on the part of the intelligent public both within and without the colleges. Thousands of the most thoughtful students all over the United States and Canada are now eager to be admitted to courses dealing with various phases of evolutionary biology. The demand for such courses has doubled or tripled within the last two or three years. Not less significant is the fact that there has been a marked increase in the number of books on evolutionary subjects, several of which rank among the best sellers.

Strange as it may seem, the least disturbed by all this excitement have been those at whom the attack has been especially aimed: the evolutionists themselves, especially those who are engaged in the teaching of evolutionary biology, or who are conducting research in the field. These men are, of all people, in the best position to judge of the validity of the principle of evolution, for they are almost the only ones with sufficient knowledge and training in the subject to justify a well-founded opinion. Certainly not less than ninety-nine per cent of these accept the broad principle of evolution and consider it to be adequately established as a law of nature. Their differences of opinion are not at all concerned with the fact of evolution, but with a far more difficult and purely technical matter: the exact causes and modes of evolution.

The opponents of evolution mistake the present disturbed and somewhat pessimistic state of opinion on the part of the experts as to the causo-mechanics of evolution for a loss of confidence in the principle itself. Nothing could be farther from the fact. A parallel situation will serve to reveal the untenability of the above position. At the present time descriptive embryology is a mature science. We have detailed information as to development of a large number of different organisms from the egg to the adult stages. Not only the fact of individual development in general has been established and can be demonstrated even to a child, but we have a great deal of technical information as to the modes and mechanics involved in growth and differentiation. In spite of all this knowledge of observed phenomena, we are almost totally in the dark as to the real causes underlying development. We do not know anything about the motive power of development nor why any particular adult form is assumed. We are as yet unable to determine just what rôle the environment plays in development, nor yet the exact mechanism of heredity. These and other technical difficulties have absolutely no bearing on the plain facts of development. In exactly the same way we are justified in claiming that the fact of evolution is in no sense weakened because advancing knowledge has revealed to us some elusive and intricate features associated with the causo-mechanical explanations of the facts.

Unquestionably the leading source of misunderstanding on the part of the enemies of evolution has been a failure

THE BULTEL-TELLIER SAND AND GRAVEL PIT AT SAINT-ACHEUL,
NEAR AMIENS, FRANCE.

This excavation is in the third terrace of the Somme valley; elevation above
sea level, about 42 meters (138 feet). The section represents eight industrial
levels, of which the sequence is as follows: L, gravels with Chellean imple-
ments, resting on chalk; F, ancient sandy loess with Acheulian workshops at
two levels; E, ancient loess with large concretions; D, brick earth of the
ancient loess with Upper Acheulian lanceolate types near the top; B', lower
part of recent loess, with Lower Mousterian industry, including many cleavers,
at level C''; B, upper part of recent loess with Upper Mousterian industry at
level c; A, brick earth with Solutrean industry about midway and Neolithic
near the surface. The weathered portion (D) of the ancient loess, known as
limon rouge, has a considerable thickness, indicating that a long period
elapsed between the deposition of the ancient loess and that of the recent loess
Photograph by Commont.

to appreciate that at present the term Darwinism is used in two very different senses: the older and better known sense in which Darwinism and the general theory of organic evolution are synonymous, and the modern technical sense in which Darwinism applies only to Darwin's own particular causal theory of the origin of species and of adaptation, a theory known as Natural Selection. Opinions as to the adequacy of this theory to explain in causal terms the phenomena it purports to explain are almost as numerous as are scientists who claim a right to such opinions. Some attach even more importance to this theory than did Darwin himself, while others hold that it has been totally discredited. The majority of biologists entertain shades of opinion somewhere between these two extremes. Some of the leading opponents of evolution, with what I am forced to believe amounts to disingenuousness, refuse to see that there is a sharp distinction between Darwinism used as a synonym for the Principle of Evolution and Darwinism used in the narrower technical sense as merely the mooted theory of Natural Selection. Taking advantage of the dual use of the term Darwinism they make the unqualified claim that most of the leading evolutionists of the present time have lost confidence in their theory, and are on the point of abandoning it. Unless they hold that the end justifies the means, it is difficult to understand how honest men could allow themselves to descend to so obvious a form of verbal trickery. The plain truth of the matter is that never before in the history of science has there been so nearly a unanimous acceptance of the Principle of Evolution and so little consensus of opinion as to its causes.

Strange as it may seem, biologists long ago ceased to concern themselves with the question of the fact of evolution, and for a long time have been devoting their energies to the solution of the extremely difficult problem of its causes. The opponents of evolution, on the contrary, have neither the training nor the equipment to tackle the problem of causes and therefore confine their attention to the question of the validity of the principle. The crux of the whole present controversy then is exactly the same as it was sixty-five years ago when Darwin published his epoch-making volume: is evolution merely a wild guess or is it an established principle supported by an adequate array of facts? The anti-evolutionists claim that the theory of evolution is based merely on assumptions, conjectures, interpretations, and false statements. If this is the situation they are wise to cling to the only alternative explanation of nature as it is: that all species of animals and plants known to-day were created in their present form, and have remained unchanged. Let us not mince matters in this connection: plants, animals, and man are either fixed and immutable products of special creation or else they have undergone changes and are undergoing changes to-day. If we admit that species have changed, and are changing at the present time, that is all that the Principle of Evolution implies. The evolutionist stands for and believes in a changing world, and unless you, the reader, believe in a fixed and unchanging world you, too, are an evolutionist. Evolution is merely the philosophy of change as opposed to the philosophy of fixity and unchangeability. One must chose between these alternate philosophies, for there is no intermediate position; once admit a changing world and you admit the essence of evolution. The particular courses of change or the causes of any particular kinds of change are matters that the expert alone is in a position intelligently to discuss. We know with certainty some few things about the course of evolution, and we believe that we have discovered some important phases of the mechanism of evolution, but these are controversial matters and in no way affect the question as to the validity of the principle. Whether or not evolution may lay claim to rank as a law of nature depends upon the strength, the coherency, and the abundance of the so-called evidences of evolution. The presentation of the evidences of evolution is the task assigned me by the editor of the present volume, and I feel impelled to hurry on toward the accomplishment of this task, but before I do so I wish to disclaim any illusions as to the possibility of convincing any special creationist of the reasonableness of the evolutionary explanation of nature. To appreciate the force of the evidences of evolution requires long and arduous study. This being the case, the only persons who have an adequate foundation for their belief in the principle are those who have made the subject a life study. The more intensively one pursues any given line of biological specialization the more necessary does it become to adopt an evolutionary point of view. The layman can obtain at best only a faint idea of the real strength of the factual basis of evolution. If this mere smattering of information seems to him conclusive, how much more conclusive must be the great masses of evidence at the command of the leading experts in the numerous fields of biological science the integrity of which depends upon the validity of the evolutionary principle. A timely warning to the lay reader of the present chapter may not be amiss. Don't expect too much in the way of cogent proof of the fact of evolution. If you earnestly desire to know for yourself how strongly founded upon fact evolution really is, there is no royal road to the object of your desire. You must become a student of biology and allied sciences. Little wonder then that leading opponents of evolution fail to see the force of the evidences of a changeable and changing world, for they never take the trouble to acquire a working knowledge of the sciences that furnish the evidence. Ignorance breeds misunderstanding and without understanding there can be no basis for intelligent controversy.

One of the leading accusations of the anti-evolutionist is that teachers of the subject fail to discriminate between what is fact, and what is mere interpretation of fact; that they teach theories as though they were facts; that they assume the principle of evolution and distort the facts into a semblance of agreement with the theory. This is a severe indictment and doubtless founded on fact to some extent, for there are all too many inadequately trained teachers of biology who, because of the somewhat sensational implications involved, attempt to teach evolutionary science. If I am at all competent to speak for my colleagues, I may say quite frankly that this form of abuse is as objectionable to the trained evolutionist as it is to the professional anti-evolutionist. On this point we are in full agreement; but the evolutionist would rather attempt to mitigate the abuse by education than to make use of our legislative machinery in a field in which it was never intended to operate. There is much poor teaching in all branches of knowledge, but to pass laws making poor teaching a crime would be no more absurd than are some of the more or less successful efforts of the anti-evolutionists to outlaw the teaching of evolution.

What the reading public needs to-day is an entirely frank and unprejudiced statement of the facts upon which the principle of evolution rests, a statement in which facts are stated as facts, inferences as inferences, and theories as theories. This kind of statement shall be our aim, and we hope to avoid any wide departures from the mark.

THE NATURE OF THE PROOF OF ORGANIC EVOLUTION

There are two distinct types of evidences of evolution, one of which has to do with changes that have occurred during past ages, the other with changes that are going on at the present time. The evidences of changes that have taken place in the remote past must in their very nature be indirect and to some extent circumstantial, for there are no living eye-witnesses of events so far removed from the present and there are no documentary records written in human language. Records of past events are written, however, for him who has learned the language, in the rocks, in the anatomical details of modern species, in the development of animals and plants, in their associations with one another and with the environment, in their classification, and in their geographic distribution past and present. Evidences that species are changing to-day are quite direct in character, for more or less radical hereditary changes have been seen in the act of taking place, though as yet we have little knowledge of the causes responsible for them. The discovery that species are changing at a noticeable rate at the present time is in itself strong evidence that they have changed in the past, and doubtless in the same ways and at the same rates of speed as those observable to-day; for even the convinced special creationist would hardly claim that species have remained immutable since their creation only to begin to change during the present era. Little can be learned about the large changes involved in organic evolution by observing the relatively small changes of the present, for it takes immense periods of time for the larger waves of change to run their course and reach their culmination. For the study of past evolutionary events we use the historical method so successfully employed in archæology and ancient history; for the study of present evolution we make use of the methods of direct observation and experiment. The findings in one field strongly support and supplement the other.

When we admit that the evidences of past evolution are indirect and circumstantial, we should hasten to add that the same is true of all other great scientific generalizations. The evidences upon which the Law of Gravity are based are no less indirect than are those supporting the Principle of Evolution. Like all other great scientific generalizations, the Law of Gravity has acquired its validity through its ability to explain, unify, and rationalize many observed facts of physical nature. If certain facts entirely out of accord with the Law of Gravity were to come to light, physicists would be forced either to modify the statement of the law so as to bring it into harmony with the newly-discovered facts or else to substitute a new law capable of meeting the situation. Laws of nature are no more or less than condensed statements about the facts of nature and therefore are valid only in so far as they agree with the facts. The Nebular Hypothesis and its modern rival, the Planetesimal Hypothesis, are both deductions from facts; they both seem to agree with many of the observed data, but neither of them is as yet fully adequate to account for all. In the field of physical chemistry we had first the Molecular Theory, then the Atomic Theory, then the Ionic Theory, and now the Electron Theory; each of these has appeared in direct response to the necessity of explaining new sets of facts, and none of them is so well founded as is the Theory of Evolution. No one has ever seen a molecule, an atom, an ion, or an electron: the existence of and the properties of these entities have been deduced from the behaviors of various chemical substances when subjected to experimental conditions.

The laws and theories just mentioned are chief among the great guiding principles of science and have gained a hold so secure that even such ultra-conservatives as the anti-evolutionists have found no occasion to object to them. For reasons which they themselves know best they choose to focus their attack upon the Principle of Evolution and upon this alone in spite of the fact that any of these other principles are as contrary to their avowed beliefs as is their pet aversion.

The Principle of Evolution stands in the first rank

among natural laws not only in its range of applicability, but in the degree of its validity, the extent to which it may lay claim to rank as an established law. It is the one great law of life. It depends for its validity, not upon conjecture or philosophy, but upon exactly the same sorts of evidence as do other laws of nature.

Now this is the test of the truth of any scientific generalization, that, when various large groups of facts that lie within the range of its application are scrutinized in the light of this generalization, the facts gain new significance, arrange themselves in systematic order, acquire coherency and unity. These services to scientific advance, and many more than these, have been rendered by the Principle of Evolution. For this reason alone, if for no other, the principle has earned a right to be respected. But in addition to all this, it must not be forgotten that evolution has been tried and tested in every conceivable way for considerably over half a century. Vast numbers of biological facts have been examined in the light of this principle and without a single exception they have been entirely compatible with it. Think what a sensation in the scientific world might be created if some one were to discover even one well-authenticated fact that could not be reconciled with the Principle of Evolution! If the enemies of evolution ever expect to make any real headway in their campaign they should devote their energies toward the discovery of such a fact.

The exact nature of the proof of the Principle of Evolution is that when great masses of scientific data such as are involved in those branches of biology known as taxonomy, comparative anatomy, embryology, serology, paleontology, and geographic distribution, are looked upon as the result of evolutionary processes, they take on orderliness, reasonableness, unity, and coherency. Not only this, but each subscience becomes more closely linked with the others and all turn out to be but different aspects of the one great process. No other explanation of biological phenomena that in any sense rivals the evolution principle has ever been offered to the public. This principle cannot be abandoned until one more satisfactory comes forth to take its place. To revert to the thoroughly discredited and unscientific idea of special creation would be as utterly impossible as to revert to the ancient geocentric conception of the universe, according to which a flat earth was thought to occupy the center of the universe and the sun, moon, and stars to revolve about it.

Let us reiterate that a theory or a principle is acceptable only so long as it accords with the facts already known and leads to the discovery of new facts and principles. Whether or not the Principle of Evolution meets these requirements the reader must judge for himself after a perusal of the facts that lie at the basis of the principle.

The evidences of evolution that we shall investigate are contained within the following fields of biology:

1. Comparative anatomy or morphology, the science of structure.
2. Taxonomy, the science of classification.
3. Serology, the science of blood tests.
4. Embryology, the science of development.
5. Paleontology, the science of extinct life.
6. Geographic distribution, the study of the horizontal distribution of species upon the earth's surface.
7. Genetics, the analytic and experimental study of evolutionary processes going on to-day.

The Fundamental Assumption Underlying the Evidences of Evolution

A careful study of the situation reveals that the entire fabric of evolutionary evidences is woven about a single broad assumption:—*that fundamental structural resemblance signifies blood relationship; that, generally speaking, the closeness of structural resemblance runs essentially parallel with closeness of kinship.* Most biologists would say that this may once have been only an assumption, but that it is now so amply supported by facts that it has become axiomatic. However obvious the validity of this assumption may be, it is the plain duty of one who attempts to justify the evolutionary principle to avoid taking steps that are in the least open to serious criticism. If we cannot rely upon this principle we can make no sure progress toward the proof of evolution.

The assumption we are now discussing is tantamount to an affirmation of the principle of heredity: that like tends to produce like. We continually employ this principle in everyday life. We fully expect the offspring of sparrows to be sparrows, of robins to be robins; and if we should ever find an instance to the contrary, we would be greatly surprised and shocked. Furthermore, we have learned by experience that offspring not only belong to the same species as the parents, but resemble the parents more closely than they do other people. Whenever we see two people whose resemblance is closer than usual we immediately come to the conclusion that such persons are relations, probably offspring of the same parents. Every one has had the experience of meeting two persons so strikingly alike that it is almost impossible to distinguish them apart, and the natural assumption is that such persons are duplicate or identical twins. Twins of this sort are vastly more closely related than are brothers or sisters, or even than are fraternal twins who are usually no more alike than are brothers and sisters of closely similar ages. It is practically established that duplicate twins are products of the early division of a single germ cell. No closer degree of kinship can well be imagined than this, for the two individuals bear the same relationship to each other as do the two bilateral halves of one body.

The writer has had an exceptional opportunity to determine the exact degrees of resemblance existing between separate offspring derived from a single egg. It so happens that a peculiar species of mammal, the Nine-banded Armadillo, almost always gives birth to four young at a time. These quadruplets are invariably all of the same sex in a litter and are nearly identical in their anatomical details. A study of their embryonic history has proven beyond question that in every case the four embryos are produced by the division of a single normally fertilized egg. Large numbers of advanced sets of quadruplet fetuses were studied statistically with the idea of determining the exact degree of their resemblance. An average of a considerable number of determinations revealed the somewhat startling fact that their coefficient of correlation is .93, which is merely another way of saying that they are 93 per cent identical. The remarkable closeness of this resemblance may be fully appreciated when it is realized that the only structural resemblances belonging to this order of closeness are those existing between the right and left halves of single individuals, and that the next order of resemblance is that between siblings (brothers or sisters), who are only 50 per cent identical.

This then is a crucial test of the validity of the assumption that closeness of resemblance is a function of closeness of kinship, for here we have the closest approach to identity in connection with what is also the closest possible blood relationship.

Employing the principle of heredity in a somewhat broader way, and in a way that is hardly likely to be questioned even by the most captious, we account for the common possession of certain structural peculiarities by all members of a given kind or species of animal by saying that characters have been derived from a common ancestor. It is only a short step in logic to conclude that two closely similar kinds or species of animal have been derived one from the other or from a common species. Once having taken this step we are on the road that leads inevitably to an evolutionary interpretation of natural groups. If the principle of heredity holds for fraternities,[2] for races, for species, where are we to draw the line? It does not seem reasonable to admit that structural resemblances within the fraternity, the race, the species, are accounted for as a product of heredity, and to deny that equally plain resemblances among the species of a genus or among the genera of a family have a hereditary basis. It is logically impossible to draw the line at any level of organic classification, and say that fundamental structural resemblance is the product of heredity up to such and such a level, but that beyond some arbitrarily settled point heredity ceases to operate.

The principle of heredity and its necessary implications constitute what we may term the fundamental assumption underlying descent with modification, or in other words, organic evolution. The special creationist assumes much more potency for heredity than does the evolutionist, for he believes in descent without modification, a sort of stereotyped heredity slavishly duplicating a fixed set of structural patterns without variation or improvement. Since both evolutionist and creationist find it equally necessary to assume the principle of heredity there should be little argument on this score. But let the reader beware at this point in the discussion, for if he admits the postulates already presented he cannot avoid the inevitable conclusion that evolution is the only adequate explanation of things as they are.

---

[2] Offspring of the same parents.

EVIDENCES FROM COMPARATIVE ANATOMY

The foundation stones of comparative anatomy are the principles of homology and of analogy. The former implies heredity and the latter variation.

The Principle of Homology

Any one who has at all seriously studied comparative anatomy must have been impressed with the fact that the animal kingdom exhibits several distinct main types of architecture, each of which characterizes one of the grand divisions of the kingdom. Within each of these great assemblages of animals characterized by a common plan of organization there are almost innumerable structural diversities within the scope of the fundamental plan. These major or minor departures from the ideally generalized condition remind one of variations upon a theme in music: no matter how elaborate the variation may be, the skilled musician recognizes the common theme running through it all. This fundamental unity amidst minor diversity of form or of function is looked upon as a common inheritance from a more or less remote ancestor. In animals belonging to the same group and therefore having the same general plan of organization we find many organs having the same embryonic origin and the same general relations to other structures, but with vastly different superficial appearance and playing quite diverse functional rôles. Such structures are said to be *homologous*.

A common example of homologous structures is presented by the fore limbs of various types of backboned animals (vertebrates): such, for example, as that of man, that of the whale, that of the bird, and that of the horse. The arm of man is by far the most generalized of these; it is not far from the ideal prototypic land vertebrate fore limb, in that it is not specialized for any particular function but is a versatile tool of the brain. The flipper of the whale is a short, broad, paddle-like structure, apparently without digits, wrist, fore arm or upper arm; but on close examination it is seen to possess all of these structures in a condition homologous almost bone for bone and muscle for muscle with those of the human arm. The wing of the bird, a highly specialized organ of flight, appears superficially to have nothing in common with the arm of man; but a study of its anatomy shows the same bony architecture and muscular complex, modified rather profoundly for a different function and with the thumb and two of the fingers greatly reduced or entirely unrepresented in the adult stage. The fore leg of the horse is a specialized cursorial appendage, and in accord with this function has but one functional toe with a heavy toe-nail or hoof. Two other toes are represented by the so-called splint bones, mere vestiges of once useful structures. In other respects the horse's leg is quite homologous with that of other land vertebrates. The evolutionary explanation of the fact that these several types of limbs (each playing an entirely different rôle in nature and each so unlike the other in form and proportions) have the same fundamental architecture, is that they have all inherited these characters from some distant common ancestor. In each case the inheritance has undergone modification in harmony with the life needs of the organism. This of course implies descent with modification, which is no more or less than evolution.

An equally significant situation comes to light in connection with the hind limbs of vertebrates. The leg of man, a specialized walking appendage, is much less versatile than is the arm; yet it is closely homologous with the latter. The hind limb of the whale is in some species entirely wanting in the adult or else is in vestigial condition. The leg of the bird is decidedly reptilian in structure and is believed to have retained in large measure the characteristics of that of the supposed reptilian ancestors. The hind limb of the horse, though somewhat stronger and heavier than the fore limb, resembles the latter closely both in form and function. Snakes are typically limbless vertebrates, but the python has small but clearly defined hind limbs, somewhat reduced in number of bones and almost entirely hidden beneath the scaly integument.

No other attempt to explain homologies such as those briefly outlined above has been made except that of special creation, and this implies a slavish adherence to a preconceived ideal plan together with capricious departures from the plan in various instances. A systematic attempt to apply the special creation concept to all cases of homologies involves one in the utmost confusion of ideas and leads almost inevitably to irreverence, which is abhorrent to evolutionist as well as to special creationist.

Vestigial Structures

These may be defined as functionless rudiments of structures whose homologues are found in a functional state in other members of a group with a common architectural plan. Thus the hind limbs of the whale and of the python, the thumb of the bird, the splint bones of the horse, are vestigial homologues of structures well developed in more generalized groups of vertebrates.

The case of the hind limb vestiges in the various species of whales may be emphasized as a crucial one. Several different degrees of rudimentation are found in different types of whales, ranging from a state in which the pelvic bones and those of most of the leg are clearly recognizable as such down to one in which these bones are entirely absent in the adult condition. In the cases where the bones are obvious, the situation is just this:—deeply buried beneath the thick cushion of blubber in the pelvic region there lies a little handful of bones, ridiculously minute in comparison with the giant proportions of the other parts of the skeleton. These bones are immovable because their muscular connections are atrophied: they do no service in supporting the frame of the animal; in short, they cannot possibly function as bones at all. The somewhat peurile argument of the anti-evolutionist that these vestigial limb bones play some useful though unknown rôle, else they would never have been created, cannot seriously be entertained in this case, for what can they make of the fact that some whales entirely lack these structures? More difficult even than this for the special creationist to explain is the fact that, even in those whales that lack vestigial limb bones in the adult condition, posterior limb buds appear in the early embryonic period and then slowly atrophy. The case just described is in no way exceptional or peculiar. It is, on the contrary, quite typical of a very general phenomenon.

Vestigial Structures in Man

There are, according to Wiedersheim, no less than 180 vestigial structures in the human body, sufficient to make of a man a veritable walking museum of antiquities. Among these are:—the vermiform appendix; the abbreviated tail with its set of caudal muscles; a complicated set of muscles homologous with those employed by other animals for moving their ears, but practically functionless in all but a very few men; a complete equipment of scalp muscles, used by other animals for erecting the hair but of very doubtful utility in man even in the rare instances when they function voluntarily; gill slits in the embryo, the homologues of which are used in aquatic respiration; miniature third eyelids (nictitating membranes), functional in all reptiles and birds, greatly reduced or vestigial in all mammals; the lanugo, a complete coating of embryonic down or hair, which disappears long before birth and can hardly serve any useful function while it lasts. These and numerous other structures of the same sort can be reasonably interpreted as evidence that man has descended from ancestors in which these organs were functional. Man has never completely lost these characters; he continues to inherit them though he no longer has any use for them. Heredity is stubborn and tenacious, clinging persistently to vestiges of all that the race has once possessed, though chiefly concerned in bringing to perfection the more recent adaptive features of the race.

Homology versus Analogy

It is quite common to find different animals with certain structures that look alike and function alike but are not homologous. The eye of the octopus, a cephalopod mollusc, has a chorion, a lens, a retina, an optic nerve, and a general aspect decidedly like that of a fish. As an optical instrument it must obviously function in the same manner as does the eye of an aquatic vertebrate; but not one part of the eye of a cephalopod is homologous with that of a vertebrate. Because these two types of eye look alike and function alike, but arise from quite different embryonic primordia adapted to meet a common function, they are known as analogous structures. They are to be sharply contrasted with homologous structures, which may be widely different in form and function so long as they arise from equivalent embryonic primordia. Both homologies and analogies imply changes in relation to the environment and therefore plainly favor the idea of descent with modification.

Connecting Links

If one group of animals has been derived by descent from another there should be some forms more or less in-

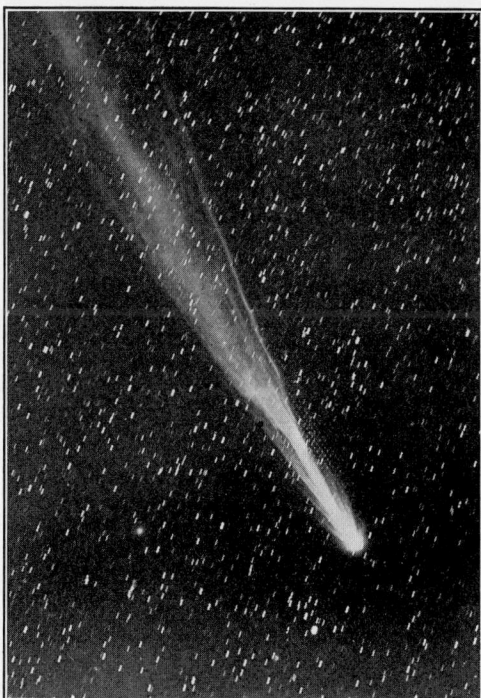

COMET MOREHOUSE.

This was photographed on November 16, 1908, by the late Professor E. E. Barnard with the Bruce telescope of the Yerkes Observatory. The tail on this date was about ten million miles long, and the comet's distance was one hundred thirty-two million miles. During the exposure of sixty-three minutes, the observer kept the telescope pointing at the comet, which caused each star to appear on the plate as a streak instead of a dot.

termediate between the two and with some characteristics
of both groups. Many such connecting links actually exist
at the present time. Almost every order of animals possesses some primitive members that have doubtless evolved
at a slower rate than their relatives and have on that account retained a larger measure of ancestral traits than
have the more typical representatives of the group. Thus
there is a group of primitive annelid worms, represented
by *Dinophilus, Protodrillus,* and *Polygordius,* that serve
partially to bridge the gap between the two grand divisions, annelids and flatworms. The case of the several
species of *Dinophilus* is especially noteworthy, for these
little animals are so evenly balanced between the characteristics of one phylum and those of the other that some
authors place them among the flatworms, others among
the annelids, and still others are inclined to place them
in an anomalous group by themselves. There is an interesting genus of primitive centipedes, called *Peripatus,*
which possesses about as many annelid features as arthro-
pod features. Among vertebrates we have the familiar example of the lung fishes with both the gills of fishes and
lungs homologous with those of land vertebrates. And
finally, we may mention those curious egg-laying mammals, momotremes, of Australia and New Zealand, which,
though obviously mammalian in most respects, possess,
in addition to laying eggs after the fashion of reptiles,
many other decidedly reptilian traits. The reader interested in following up in more detail this interesting branch
of comparative anatomy will find the subject skillfully
handled by Geoffroy Smith in a volume entitled *Primitive
Animals.*

Comparative Anatomy is a mature and well organized
science and involves a vast amount of technical data.
No one but a trained comparative anatomist can reasonably
be expected to appreciate the dependence of this subject

upon the principle of evolution. Without evolution as a guiding principle comparative anatomy would be a hopeless mass of meaningless and disconnected facts; with the aid of the principle of homology, an evolutionary assumption, it has grown to be one of the most scientific branches of biology. This may be taken as an illustration of the nature of the proof of organic evolution: that when it is used as a working hypothesis or guiding principle, it really works in that it is not only consistent with all of the facts, but lends significance and interest to facts that would otherwise be drab and disconnected.

EVIDENCES FROM CLASSIFICATION

The object of classification is to arrange all species of animals and plants in groups of various degrees of inclusiveness which shall express as closely as possible the actual degrees of relationhip existing between them. In pursuance of this object we begin by grouping together as one *species* all animals that are essentially alike in their anatomical details. As an example of the methods of classification we may take the following familiar instance:—the European wolf is a particular kind of animal constituting a species called *lupus* (the Latin word for wolf) all members of which are more like one another than they are like wolves of other sorts, for the reason that they have a common inheritance. There are not a few other species of wolves, each given a Latin name, and all of these wolf species, including dogs (believed to be domesticated and therefore highly modified wolves), are placed in one genus, *Canis*. Several other genera of more or less wolflike animals, such as jackals and foxes, are grouped with the genus *Canis*, and constitute the family *Canidæ*, the assumption being that they are all the diversified descendants of some common wolflike ancestor. Other families, such as the Cat Family (*Felidæ*), the Bear Family (*Ursidæ*), and several other families of terrestrial beasts of prey, constitute the suborder *Fissipedia*. These is turn are grouped with the marine beasts of prey, such as seals, sea-lions, walruses (suborder *Pinnipedia*) to form the mammalian order, *Carnivora*. Several other orders of animals with many characteristics in common are combined to form the class *Mammalia*, which is one of several classes belonging to the subphylum *Vertebrata*, a branch of the phylum *Chordata*. A phylum is one of the grand subdivisions of the animal kingdom and is made up of species with the same fundamental plan of organization the common features of which are believed to be derived from a common ancestral type.

The underlying assumption of classification is the same that underlies comparative anatomy: that degrees of resemblance run parallel with degrees of blood relationship, that the most nearly identical individuals are most closely related and that those that bear the least fundamental resemblance to each other are either not genetically related at all or else had a common ancestor far back in the misty past when animal life was in process of origin. We have already shown that this assumption holds good in all cases where it has been possible to put it to the test. No further justification need be offered in this place for making use of the only adequate instrument of classification: the principle of homology.

What Is a Species?

The species is the unit of classification, but there is serious doubt as to whether species have any reality outside of the minds of taxonomists. Certainly it is extremely difficult, if at all possible, exactly to draw sharp boundary lines between closely similar species. When we examine a large number of individuals belonging to a given species we find that there are no two exactly alike in all respects.

As a rule there is a wide range of diversity within the limits of the group we call a species and the extreme variants are often so unlike the type form that were it not for the intergrading steps between them they would often be adjudged distinct species. Moreover, the species of a prosperous genus are so variable that it becomes an almost impossible task to determine where one species ends and another begins, so closely do they intergrade one into another. A species, then, is not a fixed and definite assemblage such as one would expect it to be if specially created as an immutable thing. On the contrary, intensive study of any widely distributed species gives the impression of an intricate network of interrelated individuals changing in a great variety of ways.

The completed classification of any large group, such as the vertebrates, presents itself as an elaborately branching system whose resemblance to a tree is unmistakable. The phylum branches into subphyla, some of the latter into several classes, classes into orders, orders into families, families into genera, genera into species, species into varieties. We may compare the phylum to one of the main branches coming off from the trunk, while the varieties may be thought of as the terminal twigs. This tree-like arrangement is exactly what one would expect to find in a group descended from a common ancestry and modified along many different lines. It is in reality a genealogical tree. If this striking arrangement is a part of the plan of special creation it is indeed strangely unfortunate that it speaks so plainly of descent with modification.

Man's Place in the System of Classification

There is no greater difficulty in connection with the classification of man than in that of any other living species. Indeed there are scores, even hundreds, of species whose exact affinities with other groups are far less ob-

vious than those of the human species. Anatomically, the
genus *Homo* bears a striking resemblance to the anthropoid apes. Bone for bone, muscle for muscle, nerve for
nerve, and in many special details, man and the anthropoid apes are extremely similar. Homologies are so obvious that even the novice in comparative anatomy notes
them at a glance. Man is many degrees closer anatomically to the great apes than the latter are to the true
monkeys, yet the special creationist insists upon placing
man in biological isolation as a creature without affinities
to the animal world. If man is a creature apart from all
animals it is extremely difficult to understand the significance of the fact that he is constructed along lines so
closely similar to those of certain animals; that his processes of reproduction are exactly those of other animals;
that in his development he shows the closest parallelism
step for step to the apes; that his modes of nutrition,
respiration, excretion, involve the same chemical processes;
and that even his fundamental psychological processes are
of the same kind, though differing in degree of specialization, as are those of lower animals.

Comparative anatomists recognize man as a vertebrate,
for he has all of the characteristic features of that group.
He is obviously a mammal, for he complies with qualifica-
tions of that class in having hair; in giving birth to living
young after a period of uterine development; in suckling
the young by means of mammary glands; in having two
sets of teeth one succeeding the other; in having the teeth
differentiated into incisors, canines, and molars; and in
many other particulars of skeleton, muscular system, circulatory system, alimentary system, brain, and other parts
of the central nervous system. Among mammals, man
belongs to the well-defined order of Primates, an order
anatomically about halfway between the most generalized
and the most specialized of the mammalian orders. Apart

from his extraordinary nervous specialization, man is a relatively generalized mammal as compared with such highly specialized types as, for example, the whales. The older taxonomists placed man and the other primates at the top of the genealogical tree, assigning to him the central tip of the central branch as though the goal of all organic evolution were man. Accordingly, those mammals such as the whales, which are least like man, were considered the lowest members of the class. There has been within recent years a pronounced reversal of this anthropocentric point of view, which has resulted in a complete revision of the arrangement of mammalian orders, with the Insectivora the lowest, the Cetacea (whales) the highest, and the Primates about intermediate in systematic position.

The Order Primates consists of two Suborders—Lemuroidea and Anthropoidea. The lemurs or half apes are small arborial animals with somewhat squirrel-like habits but with flat nails and certain other primate characters. They serve to link up the Primates with the most primitive of the mammalian orders, the Insectivora, which are now believed, on anatomical and paleontological grounds, to be ancestral not only to the primates but to most of the other modern mammalian orders. The anthropoid or manlike Primates are divided into four distinct families: the *Hapalidæ* or marmosets; the *Cercopithecidæ* or New World monkeys; the *Simiidæ* or anthropoid apes; and the *Hominidæ* or men. The family *Hominidæ* includes four genera: The genus Pithecanthropus, represented by the fragmentary remains of an extinct Javan ape-man, the genus *Paleanthropus*, the genus *Eoanthropus*, and the genus *Homo,* including in addition to the existing species, *Homo sapiens,* several different extinct human species known as the Dawn Man, the Neanderthal Man, the Rhodesian Man, and others.

The species *Homo sapiens* consists of at least four sub-species or major varieties, each consisting of numerous minor races and admixtures of these. This high degree of diversity within the species is evidence of rapid evolution. If a little over four thousand years ago, as the special creationists claim, one man was created and has become the ancestor of all men living to-day, evolution must have gone on at an extremely rapid rate in order to have produced so many widely different races, for there could scarcely have been more than one hundred and twenty generations in that time. If species are believed to be immutable it is difficult to understand why man should be such a diversified group as he is.

EVIDENCES FROM BLOOD TESTS

The methods of classifying animals just outlined depend upon relatively gross criteria (homologies) as compared with the refinements characteristic of the serological technique used in blood testing. This latter method of classifying animals depends upon chemical similarities and differences in the bloods of various animals, and the basic assumption is once more that degrees of resemblance parallel degrees of blood relationship. Recent investigation has shown that certain materials in an animal's blood are even more sharply specific than are its visible structural characteristics. Chemical tests of extreme delicacy are used to reveal resemblances in blood. Thus, if we wish to find out what animals are most like man in blood composition we can find it out in the following manner: Human blood is drawn and allowed to clot, a process that separates the solid materials in the blood from the liquid serum. The latter watery fluid contains the specific human blood ingredients. Small doses of it are injected at two-day intervals into the blood vessels of a rabbit. At first the rabbit is sickened by the injection, thus showing a marked reaction to the foreign material. In the course of a short time, however, there is no further reaction, and we may conclude that the rabbit is immunized. What has happened is that some substance has been developed in the rabbit's blood which neutralizes the toxic effects of human blood. It is a sort of antitoxin and may be spoken of as anti-human serum, a material that may now be used as a delicate indicator of blood kinship. When this anti-human serum is mixed with serum taken from the blood of any human being an immediate and definite white precipitate is formed; when mixed with that of any of the anthropoid apes the precipitate is similar to that formed with human serum but less abundant and somewhat slower in appearing. The tests showed a less prompt and less abundant reaction with the blood of Old World monkeys, a slight but definite reaction with that of New World monkeys, and no noticeable reaction with that of lemurs.

The tests further indicated that, if strong enough solutions are used and time enough allowed for the precipitate to settle, there is an unmistakeable blood relationship among all mammals and that degrees of relationship run closely parallel with those based upon homologies. Not only this, but not a few affinities, the existence of which had been only vaguely suggested by comparative anatomy, are strongly emphasized by blood tests. One most remarkable revelation is that whales, the most specialized among mammals, are more closely related to the ungulates (hoofed animals), and especially to the swine family, than to any other group of the Class Mammalia—a diagnosis that had previously been made by several anatomists on what appeared to be rather slender morphological grounds.

At the present time the technique of blood testing for animal affinities is rather difficult and very few workers have attempted to make use of it. The results so far attained, however, are so definite and clean-cut that there is every reason to expect a great future for this new type of evolutionary evidence. Many groups of animals have already been tested and in general the affinities indicated closely parallel those based on homologies. There is, however, no exactness about this parallel; nor could we expect such to be the case. For that matter there is no exact parallelism between the teeth and the feet, between the head and the tail. No two systems of an organism exactly keep pace in their evolution: one may remain relatively conservative while the other may become greatly specialized. Of all systems, the blood appears to have been the most conservative and to have retained most fully its ancestral characters. It is on this account that blood tests are so valuable in revealing relationships that can scarcely be determined in any other way.

Far more important than any information as to animal affinities revealed by blood tests is the fact that the classification of animals based on blood tests is essentially the same as that based on morphology. Suppose, for the sake of argument, that these two modes of classification had revealed quite contrary arrangements: what a blow to our confidence in the validity of evolution! Conversely, what a strong support of the evolution principle is afforded by the fact that the two systems of classification point to the same lines of descent!

EVIDENCES FROM EMBRYOLOGY

There should be no sharp division between the evidences from Comparative Anatomy and those from Embryology. These two branches of biology are inseparable: one must be interpreted in the light of the other. Comparative anatomy deals with the adult structures of organisms. Whenever there is any question about homologies of fully developed structures recourse is had to younger and still younger stages, for when structures are really homologous they tend to be more closely similar the younger they are. Structures that come from the same or similar embryonic primordia are by definition homologous. ⟨Therefore the only certain test of homologies is a study of embryology.⟩

It is necessary to bear in mind that an individual is not merely his adult condition; that a species is not fully defined by a description of its adult characteristics. The species characteristics include those of the egg and the sperm, the cleavage pattern, the particular modes of gastrulation and of further differentiation. In brief, the species is fully defined only by a full description of its entire ontogeny. Very closely related species keep step nearly all the way through their ontogenies and diverge only toward the end of their courses. Distantly related forms diverge comparatively early in their developmental paths; while unrelated forms may have little or nothing in common from the beginning.

The most advanced groups of organisms travel a much longer journey before reaching their destination than do organisms of lower status. In many instances certain early stages in the development of an advanced organism resemble in unmistakable ways the end stages of less advanced organisms. There is, in fact, in the long ontogeny of members of higher groups, a sort of rough-and-ready repetition of the characteristic features of many lower groups. This fact has so impressed some biologists that they have embodied it into a law, the so-called biogenetic law; that ontogeny recapitulates phylogeny. In less technical language this means that the various stages in the development of the individual are like the various ancestral forms from which the species is descended, the earliest embryonic stages being like the most remote ancestors and the later stages like the more recent ancestors. In still other words, the concept may be stated as follows:

the developmental history of the individual may be regarded as an abbreviated résumé of its ancestral history.

There is serious objection to this idea not only on the part of anti-evolutionists but on that of most biologists. In the first place it is obvious that no embryonic stage can be in any real sense the equivalent of any adult ancestor. The most we can affirm is that while some embryonic characters of the higher group strongly remind us of some adult features of lower groups, the *tout ensemble* of the former is not at all closely similar to that of the latter. In the second place, it should not be forgotten that the embryonic and larval stages of organisms have much more pressing demands upon them than that of recording their ancestral attainments—they must adapt themselves to their surroundings if they are to survive. As a result of this pressing necessity many larvæ and even embryos are so profoundly modified in adaptive ways that their ancestral characters are largely obscured. Various larval or fetal organs commonly furnish the outstanding characteristics of developmental histories, and these purely temporary organs not only tell no story of ancestries but frequently so mask the ancestral story as to make it almost indecipherable. In the third place, different systems of organs develop at different rates, so that when one system has reached an advanced state of differentiation another system may be still in the primordial state. Thus, in the development of fishes the nervous system is far along its course of development before the circulatory system has even begun to differentiate. At such a stage as this the embryo is obviously not equivalent to any adult ancestor, for an organism with so discordant an organization could not survive.

In spite of its faults and limitations, however, the idea that ontogeny tends to repeat phylogeny, if used intelligently and not over-applied, is a very useful one. Organisms inherit not only their adult characters from their ancestors, but also their general developmental patterns. It is therefore inevitable that many features that have been outgrown or subordinated in modern types should be found in a state more nearly ancestral during the embryonic stages. And especially is this the case when particular systems are studied separately. Thus, we find that the human circulatory system develops through a series of stages that are much like the adult conditions of a series of ascending vertebrate classes. The heart differentiates from a sheet of mesoderm lying beneath the pharynx. It has at first the form of two nearly straight tubes, which soon fuse for part of their length to form a single tube divided at the two ends into two tubes. Later the single tube differentiates lengthwise into two cavities, the auricle and the ventricle, and is now in the stage equivalent to that of an adult fish. The auricle next divides into two chambers, thus resembling that of an amphibian. Finally the ventricle subdivides also, giving rise to the four-chambered heart characteristic of mammals. The main arteries and veins of the head region are at first laid down with reference to what are known as the branchial arches, the structural framework of the branchial or gill apparatus of aquatic vertebrates. Later, the whole architecture of this system becomes profoundly modified in adaptation for lung respiration. While the arteries and veins are in the fish-like condition there appear at the anterior end of the body in the prospective neck region four pairs of crevices, gill slits, which in fishes open directly into the pharynx and furnish a surface for gills. In the human embryo, however, these clefts never break through, but, after persisting for some time without playing any useful rôle, gradually disappear. The only persistent residue of the gill slits is the Eustacean tube, which connects the pharynx with the middle ear. Never at any time do the gill slits function in a respiratory capacity, for they never possess any branchial tissue. Only one interpretation of these transitory gill slits of man can be seriously entertained, namely, that, although these structures are inherited from the early aquatic ancestry, adaptive demands have caused their suppression in favor of more useful structures. Inheritance causes their appearance; lack of function prevents their development and causes their disappearance or modification.

Nothing is to be gained by a multiplication of parallelisms such as the above. Suffice it to say that the nervous system, the alimentary system, the urogenital system and other systems go through stages similar to those described above and that these resemble adult stages of lower classes of vertebrates. The embryology of man is now pretty thoroughly known in spite of the great difficulty of obtaining the early stages. Step for step it is almost precisely like that of other primates, especially like that of the anthropoids, and it is only in the latest stages that it takes on distinctly human characteristics. This is not equivalent to saying that the expert embryologist is in any doubt as to the diagnosis of a human embryo no matter how early the stage, for there are specific features about all embryos from the egg stage on to the end of development that may be distinguished by any one sufficiently versed in the subject. In spite of these specific differences, however, there can be no question that the embryology of man and that of any of the anthropoid apes show the closest of resemblances at every stage and diverge sharply only in the late stages of prenatal life. So close a resemblance in developmental histories is found only in species that are members of the same ancestral stock, for they have both inherited the characteristic features of their development from their common ancestors.

The evidence of human evolution as derived from a study

of embryology is in no wise exceptional; on the contrary
it is quite typical and may be taken as indicating that
from the developmental standpoint man is at one with
other animals.

In concluding this brief statement as to the evolutionary
significance of embryology, it should be said that the anti-
evolutionist has no explanation to offer for these facts. He
seems content to deny them.

EVIDENCES FROM PALEONTOLOGY

Paleontology is the science of ancient life. Its materials
are the more or less completely preserved remains of animals and plants that once lived. We call these remains
fossils. Fossils are real; they cannot be explained away.
If evolution has taken place and samples of every species
that has lived were preserved for study it would still be
a task of immense difficulty to work out the pedigrees of
all types of organisms now living, and we might still be
largely in the dark as to the causes of the observed changes.
As it is, we have fossil remains of perhaps only about one
out of each thousand extinct species, a mere random
sampling of the types that prevailed during the various
past ages. Considering how many factors have been at
work to prevent fossilization of large groups of species
and how erosion and metamorphosis have worked together
to destroy those fossils already preserved, we marvel that
our fossil record is sufficiently complete to tell any sort
of sequential story. The fact is that the record is sur-
prisingly full and rich.

Age of the Earth

One of the chief arguments against evolution has been
that there has not been time enough for such vast changes
to have taken place. This argument loses all of its force
in the face of what we now know about the age of the

Earth. According to the most recent computations based on the rate of radium emanation, 1,000,000,000 years have elapsed since the earth attained its present diameter. Various estimates as to the time since the first life appeared upon the surface of the globe range from 50,000,000 years to about ten times that figure. Even the lowest figure gives ample time for any sort of evolutionary change, no matter how slow.

The Earth's Strata as Time Markers

The crust of the earth is arranged in a series of horizontal strata of varying thickness. The lowest layers are obviously the oldest, except in a few localities where breaks and tilts have occurred. Even in the most disturbed mountainous regions it is an easy task for the geologist to determine the original order of the strata. This is not the place to enter upon a description of the modes of rock formation or the ways in which fossils are preserved. We shall simply go ahead on the assumption that the fossils of the lowest layers of rock are the oldest and those in the uppermost layers are the most recent.

The Main Facts Revealed by the Fossil Record

(1) None of the animals of the past are identical with those of the present. The nearest relationship is between a few species of the past which have been placed in the same genera as those of to-day.

(2) The animals and plants of each geologic stratum are at least generically different from those of any other stratum.

(3) The animals and plants of the oldest geologic strata represent all of the existing phyla except the vertebrates, but the representatives of the various phyla are relatively generalized as compared with modern representatives of the same phyla.

(4) There is a gradual progression toward more highly specialized forms as one proceeds from lower to higher strata.

(5) Many groups of animals reached the climax of their specialization long ages ago and have become extinct.

(6) Only the less specialized relatives of these most highly specialized types survived to become the progenitors of the modern representatives of the group.

(7) It is common to find a new group arising near the close of some geologic period when vast climatic changes were taking place. Such an incipient group almost regularly becomes the dominant group of the next period, presumably because it arose in response to the new conditions that accompanied the change from one period to another.

(8) The evolution of the vertebrate classes is more satisfactorily shown than that of any other group, probably because it arose within the period which is characterized by an abundant fossil record. Of the vertebrates, the mammals are best represented and show the most complete fossil pedigrees; this, because they are the most recent in origin and their remains have been least disturbed.

(9) Many practically ocmplete fossil pedigrees have been worked out, connecting modern specialized types with simpler and more generalized ancestors. Such pedigrees have been worked out for the horse, the elephant, the camel, the rhinoceros and other equally specialized modern types. A single example of this type of evidence will be given: that of the horse. Many other pedigrees have been worked out that are equally complete and no less significant.

Pedigree of the Horse

As recorded by Dendy, the course of evolution of the horse family (Equidæ) "has evidently been determined by the development of extensive, dry, grass-covered, open plains on the American continent. In adaptation to life on such areas structural modification has proceeded chiefly in two directions. The limbs have become greatly elongated and the foot uplifted from the ground, and thus adapted for rapid flight from pursuing enemies, while the middle digit has become more and more important and the others, together with the ulna and the fibula, have gradually disappeared or been reduced to mere vestiges. At the same time the grazing mechanism has been gradually perfected. The neck and head have become elongated so that the animal is able to reach the ground without bending its legs, and the cheek teeth have acquired complex grinding surfaces and have greatly increased in length to compensate for increased rate of wear. As in so many other groups, the evolution of these special characters has been accompanied by gradual increase in size. Thus *Eohippus,* of Lower Eocene times, appears to have been not more than eleven inches high at the shoulder, while existing horses measure about sixty-four inches, and numerous intermediate genera for the most part show regular progress in this respect.

"All these changes have taken place gradually, and a beautiful series of intermediate forms indicating the different stages from *Eohippus* to the modern horse have been discovered. The sequence of these stages in geological time exactly fits in with the theory that each one has been derived from the next below it by more perfect adaptation to the conditions of life. Numerous genera have been described, but it is not necessary to mention more than a few."

The first indisputably horse-like animal appears to have been *Hyracotherium* of the Lower Eocene of Europe. Another Lower Eocene genus is *Eohippus,* which lived in North America, probably having migrated across from Asia by the Alaskan land connection. In *Eohippus* the forefoot had four hoofed toes of nearly equal size, the homologue of the thumb having been reduced to a vestige. In the hind foot the great toe had entirely disappeared and the little toe had been reduced to a splint bone. Then came *Orohippus* of the Upper Eocene, *Mesohippus* of the Lower Miocene, *Protohippus* of the Lower Pliocene, *Pliohippus* of the Upper Pliocene, and finally, *Equus* of the Quarternary and Recent. This history, in so far as it concerns the characters already described, furnishes all of the intermediate conditions and perfectly connects the horses of the past with those of the present. One could hardly ask for a clearer or more conclusive story of evolution than this, and this is only one of many similar cases.

The Fossil Pedigree of Man

There is nothing peculiar or exceptional about the fossil record of man. It is considerably less complete than that of the horse, the camel, the elephant, and other purely terrestrial mammals, but it is far more complete than that of birds, bats, and several types of arboreal mammals. Much has been said by the anti-evolutionists about the fragmentary nature of the fossil record of man, but many other animals have left traces far less readily deciphered and reconstructed.

The outstanding fact brought out by a study of human paleontology is that of man's antiquity. According to the Mosaic account of the creation man appeared upon the earth somewhere between four thousand and six thousand years ago. According to the most expert testimony available, the oldest fossil in the human series is about half a million years old; and even this estimate makes man a recent product of evolution as compared with many contemporaneous animals. The earliest fossil remains of the present species of man (*Homo sapiens*) have been very conservatively estimated as 25,000 years old, while other species of extinct man date back to a period at least 100,000 years

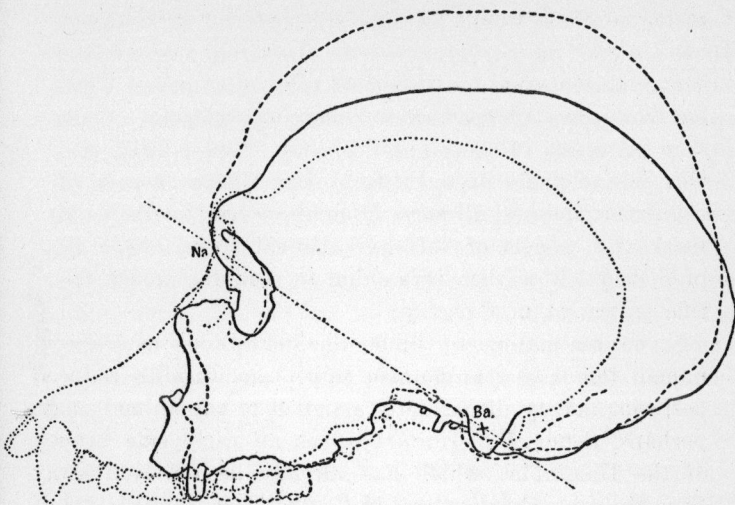

STAGES IN CRANIAL DEVELOPMENT.

Profiles of the crania of a chimpanzee (dotted line), the Mousterian of Chapelle-aux-Saints (solid line), and a modern Frenchman
(dashed line). All are placed on a common basi-nasal line of equal
length in each case (*Na.-Ba.*). They especially emphasize the differences in length of muzzle and in cranial height. After Boule.

ago. In addition to several species of the genus Homo,
anthropologists distinguish three other genera of the man
family (Hominidæ): *Pithecanthropus, Paleanthropus,* and
Eoanthropus, all more primitive than any members of the
genus *Homo.* A brief, but frank, statement about each of
these links in the human pedigree is all that is necessary
for our purposes.

Pithecanthropus erectus

This is the so-called Java Man, formerly called the Ape
Man or Missing Link, but now adjudged to be definitely
human. The fossil remains consist of a complete calvarium
or skull cap, three teeth, and a left thigh bone. These were

scattered over twenty yards of space and were discovered at different times. There is no proof that these remains belong to the same individual or even to the same species, but they are all human in their anatomical characters and they occurred in fossil-bearing rock about 500,000 years old. Many pages of scientific romance have been written about this species; all sorts of more or less justifiable pictures and models of this hypothetical species have been published. It is then refreshing to read the coldly scientific statement of Gregory:

"The association of gibbonlike, skull-top, modernized human femur and subhuman upper molars with reduced posterior moiety, if correctly assigned to one animal, may, perhaps, define *Pithecanthropus* as an early side branch of the Hominidæ, which had already been driven away from the center of dispersal in Central Asia, by pressure of higher races. But whatever its precise systematic and phylogenetic position, *Pithecanthropus,* or even its constituent parts, the skull-top, the femur and the molars, severally and collectively testify to the close relationship of the late Tertiary anthropoids with the Pleistocene Hominidæ."

Some question as to the authenticity of the published account of the remains of *Pithecanthropus* arises out of the fact that their custodian, Doctor Dubois, will not permit its further study by his colleagues.[1] The extreme fragility of these valuable relics is perhaps sufficient extenuation for what might appear to be a selfish attitude.

Paleanthropus heidelbergensis

This genus and species, commonly known as the Heidelberg Man, is based solely upon a single lower jaw in an

[1] Since this was written news has reached us that at least one leading American anthropologist has been permitted to examine the *Pithecanthropus* remains; but no report has been made public.

excellent state of preservation, with all teeth in place. The strong points about this find are, first, that it was found in a stratum whose age had been well established; and second, that its discoverer ranks among the leading experts in the field. The age of this venerable relic has been determined as at least 400,000 years, a little more recent than *Pithecanthropus*. The jaw is very primitive, heavy, and clumsily constructed as compared with that of modern man. It lacks the chin prominence, as does the jaw of the gorilla. The teeth are strictly human, though rather larger than those of modern man. This apelike jaw with human teeth forms an authentic link in the series connecting man with the anthropoids.

Eoanthropus dawsonis

This most ancient English human relic has been called the Dawn Man of Piltdown. Owing to the fact that the skull fragments had been badly damaged and scattered by workmen before they came into scientific hands, there has been a great deal of controversy as to their significance. Until the experts arrive at an agreement about this type it might be well for others to reserve judgment. There can be no doubt as to the fact that these remains show a curious admixture of simian and human characteristics, the jaw and teeth being even more simian than that of the Heidelberg Man, while the skull, though primitive, is distinctly human. The age of the Dawn Man is placed at about 200,000 to 300,000 years.

In striking contrast with the fragmentary character of the remains just described are those of three distinct species of the genus *Homo*, which are now to be briefly characterized.

Homo neanderthalensis

The well-established race known as Neanderthal Man is represented by many individual skeletons of varying degrees of completeness and showing a considerable range of diversity. Specimens have been found in France, Spain, Belgium, Germany, and Austria. This species of primitive man was of low stature, about five feet three inches in the males and less in the females. The posture was somewhat stooping. The relatively large head was long and flat, with apelike brow ridges and scarcely any forehead, and was borne on an immensely muscular neck in such a way that the face was thrust forward in simian fashion. The lower jaw was heavy and lacked a chin prominence. The teeth were of a type known as taurodont, adapted to a coarse vegetable diet and quite different in structure from those of modern man. The brain of this ancient *homoneanderthalensis* was large and specialized in some parts, but deficient in those parts associated with the higher mental functions.

There can be no question that Neanderthal Man was much more primitive, more simian in organization, than modern man. Expert opinion, as expressed by Kieth, looks upon him as "a separate and peculiar species of man which died out during or soon after the Mousterian period." This dates him back to about 50,000 years ago.

Homo rhodesiensis

Rhodesian Man is represented by a perfect skull and a nearly perfect lower jaw, the tibia, both ends of a femur, collar bone and parts of the scapula and pelvis. Part of the upper jaw of a second specimen was found in the same locality, the Broken Hill mine in northern Rhodesia. This species is largely of technical interest, and need not be described in detail. Suffice it to say that in some respects it was as primitive as Neanderthal Man, but in other respects showed distinct tendencies toward the modern condition. Anthropologists have as yet not reached a decision as to the exact taxonomic status of Rhodesian Man, nor has its age been definitely determined.

Homo sapiens

The earliest fossil evidences of the existence of our own species date back to about 25,000 years ago. At that time there lived a remarkable race, known to us as Crô-Magnons, a race said to be the most perfect physically of which we have any knowledge. Five essentially complete skeletons form the basis of the type description. This tall, strong, obviously intelligent, and artistic race, was different in several important particulars from any modern race. A detailed description of his characteristics would take us too far afield. Our chief interest in this race is that it serves to emphasize the antiquity of our own species.

In conclusion it may be said that the fossil evidences of man's ancestry are neither rich nor poor; that anthropology is a comparatively youthful science; and that new discoveries in the field are being made at a very satisfactory rate.

Evidences from Geographic Distribution

Just as paleontology deals with the vertical distribution or distribution in time of species, so geographic distribution deals with their horizontal distribution upon the earth's surface at any given period of time. Geographic distribution is a sort of cross-section of vertical distribution, giving a picture of the complex evolution of organisms at a given moment in the process. Explorers and collectors have amassed a vast amount of data as to the present and past ranges of animals and have mapped out the distribution of the majority of known species. A composite map of the geographic distribution of all known species would be the most intricate picture puzzle imaginable, and it would be almost impossible to make sense of it. A study of the distribution of limited groups, however, should lead to some reasonable explanation of their interrelations. Obviously animals are not distributed strictly according to climatic conditions or habitat complexes, for a given climate in one part of the world is associated with an entirely different fauna from a practically identical climate in another part of the world. Moreover, animals are not always or even very frequently located in those parts of the world that would offer them the best possible life conditions. This is borne out by the fact that not a few animals, when taken out of the normal range and transferred to a distant region, thrive much better than in their native territory. Thus European rabbits, when carried to Australia, throve and multiplied beyond all expectation till they became a pest. Again, as may be easily observed, the English sparrow seems to find America much more congenial than the British Isles.

If animals are not distributed according to habitats, how, then, can we account for their distribution? It is not at all likely that species retain the same ranges for long periods; they are continually changing their locations. We know, also, that the likeliest places to look for two closely similar species are adjacent territories, separated by geographic barriers. A study of the distribution of the species of a large genus usually reveals the fact that the most generalized or type species occupies the central part of the area and that the most specialized species occupy outlying areas adjacent to or connected with the main range of the genus. Taking these and related facts into consideration, we are able to offer as an explanation of the distribution of groups of allied species that a parent species originates in one place, multiplies, and tends to migrate centrifugally in all directions, modifying as it goes to fit new conditions. Some of the extreme migrants become isolated from the main body of the species and, no longer

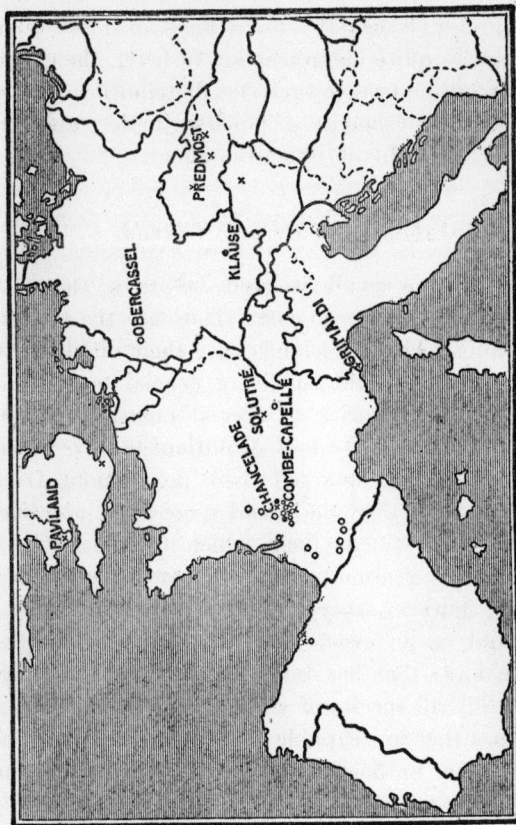

MAP SHOWING THE GEOGRAPHIC DISTRIBUTION IN EUROPE OF UPPER PALEOLITHIC
HUMAN SKELETAL REMAINS.

Aurignacian discoveries are indicated by crosses; Solutrean, by black circles;
and Magdalenian, by white circles.

interbreeding with them, become at first well-marked local
varieties and in time new species. The above is the usual
hypothesis employed in explaining geographic distribution,
and it obviously implies evolution. When used as a means
of unraveling the intricate tangle involved in the distribution of species, it has thrown a flood of light upon
situations otherwise quite inexplicable. In brief, the evolution hypothesis rationalizes geographic distribution, makes
a science of what was formerly a hopeless jumble, and has
thus proven itself a valuable scientific agent.

The Inhabitants of Oceanic Islands

Oceanic islands are small isolated bodies of land of
volcanic origin, far from continents. They are the tops of
oceanic mountains. All such islands have their inhabitants,
and a study of these should furnish a crucial test of the
validity of the rival theories of special creation and of
evolution. Both creationists and evolutionists agree that
these islands must have obtained their populations from
continental bodies. If then the island species are identical
with those of the continent from which they have been
derived, there is no reason to believe that evolution has
taken place; if, however, they are different, the degree of
difference should be an exact measure of the amount of
evolutionary change that has taken place. What are the
facts? Practically all species of animals inhabiting oceanic
islands are types that are capable of transportation in the
air during storms or on floating débris. All species belong
to the faunistic groups characteristic of the most available
continent, but the species are for the most part peculiar,
that is, different from species anywhere else. They may
belong to the same genus or family as do those of the
continent, but they are at least specifically, frequently
generically, different from the latter. Such being the case,

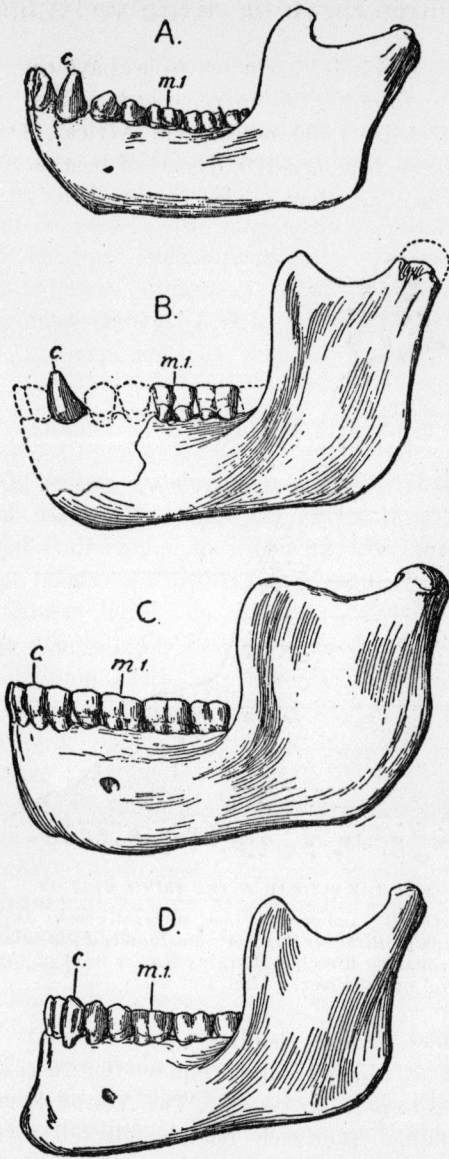

STAGES IN THE DEVELOPMENT OF THE LOWER JAW.

Outer side of the lower jaw of Piltdown man (B), compared with that of chimpanzee (A), Heidelberg man (C), and modern man (D). c, Canine tooth; m.1, first molar. Scale, ½. After Smith Woodward.

we are forced to conclude that new species have originated under island conditions. The extreme case is that of the island of St. Helena, 1,100 miles from Africa. On this little body of land there are 129 species of beetles, all but one of which are peculiar. The species belong to 39

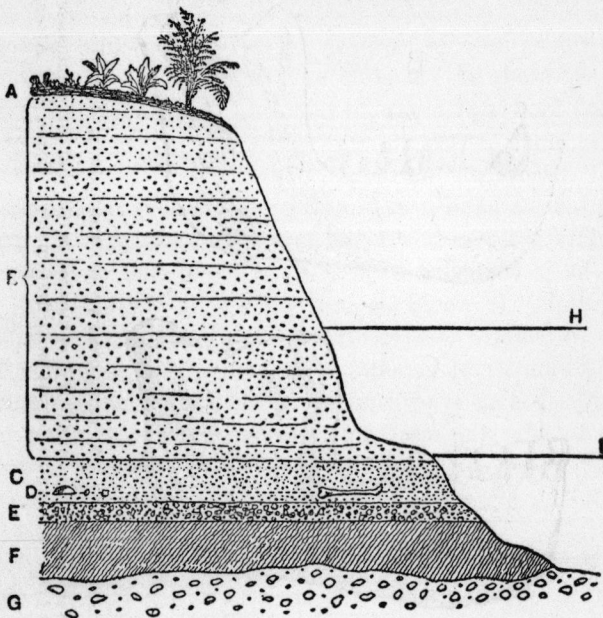

SECTION OF THE DEPOSIT AT THE TRINIL STATION.

A, vegetal earth; B, sand-rock; C, bed of lapilli-rock; D, level in which the remains of *Pithecanthropus* were found; E, conglomerate; F, clay-rock; G, marine breccia; H, rainy-season level of the river; I, dry-season level of the river. After Dubois.

genera, of which 25 are peculiar. There are 20 species of land snails, of which 17 are peculiar. Of 26 species of ferns 17 belong to peculiar genera. The Azores, Bermudas, Galapagos Islands, Sandwich Islands, all tell much the same story, but their populations are not quite so peculiar.

Limitations of space forbid the further presentation of numerous other bodies of data concerning geographic distribution, much of which testifies just as strongly for evolution as that just presented.

EVIDENCES FROM GENETICS

Genetics may be defined as the experimental and analytical study of Variation and Heredity, the two primary causal factors of organic evolution. As such, genetics aim not so much at furnishing evidence of the *fact* of evolution as at discovering its *causes*. Incidentally, however, when man takes a hand in controlling evolutionary processes and actually observes new hereditary types taking origin from old, he is observing at first hand the actual processes of evolution. It would be impossible with the scope of the present chapter to present even an adequate outline of the subject-matter of genetics. We shall merely say that the geneticist is an eye-witness of present-day evolution and is able to offer the most direct evidence that evolution is a fact.

SUMMARY OF EVIDENCES

All of the lines of evidence presented point strongly to organic evolution, and none are contrary to this principle. Most of the facts, moreover, are utterly incompatible with the only rival explanation, special creation. Not only do these evidences tell a straightforward story of evolution, but each one is entirely consistent with all of the others. Furthermore, each line of evidence aids in an understanding of the others. Thus embryology greatly illuminates comparative anatomy and classification; geographic distribution is aided by paleontology, and *vice versa;* blood tests and classification throw mutual light the one upon the other. The evolution principle is thus a great unifying and integrating scientific conception. Any conception that is so far-reaching, so consistent, and that has led to so much advance in the understanding of nature, is at least an extremely valuable idea and one not lightly to be cast aside in case it fails to agree with one's prejudices.

CHAPTER IX

SOCIAL EVOLUTION

By·Ellsworth Faris [1]

WHEN Rasmusson returned to Etah after a journey to the north of Greenland, he heard from
the Eskimos the news of the World War. "And
the fighting still goes on," they told him, "and the white
men are all killing each other. It may be that the ships
will come no more to the *Land of Men*." The Eskimos regard themselves as distinctly superior to the men of any
other race. So, also, do the Bantus, the Maoris, the
Melanesians, the Todas, the Chinese, Germans, Americans,
and Nordics. If we, then, being civilized know what it
is to be ethnocentric, how much more shall we be on our
guard when we try to maintain a scientific attitude toward
the question of the course of human development which
has lasted just to this present time and which seems to
have converged upon us as a goal.

Social Evolution is a difficult subject to discuss without
bias for it is often used as a synonym for Social Progress
to which it is indeed closely related. Like immortality and
democracy, progress is believed in because it is desired.
While it refers primarily to the past, it cannot be unmind-
ful of the future; it is at once a record and a prophecy, or
at least a hope.

---

[1] Professor of Sociology, University of Chicago.

PRIMITIVE MAN

Social evolution cannot be discussed without a discussion of primitive man, and primitive man was dead and gone long before any one ever seriously discussed anything. And since primitive man could not be found when the discussion started, he had to be invented. In the mythologies of all races he may be found, but the fantastic records have chiefly a literary value. Of course mythology furnishes certain indirect evidence concerning the mental and emotional life of a people, but we treat the material as illustrating the wishes, nothing more. But not only in the myths did this invented primitive man have an imagined existence, for in the seventeenth century he became a scientific hypothesis, being described as gentle and innocent in the books of Rousseau, cruel and selfish in the books of Hobbes, quite unformed in the books of Locke, while he is quite worthless to us in the books of them all.

Scientific study of primitive man got a bad start, for it took a false lead. In the nineteenth century primitive man was supposed to exist in the static and congealed cultures of those uncivilized peoples such as the natives of Australia, Central Africa and Melanesia. It took a long time and involved a great waste of effort before it finally became clear to all that none of these peoples are primitive, for their culture is a real culture and is very old, their languages are rich and complex, their blood is everywhere mixed, and real primitive man must be sought elsewhere than among peoples now existing.

And then they dug for him. What little we do know about him is the result of the work of the archeologist whose patient effort has built up a structure which gives us a picture of what took place in northwestern Europe, but leaves all the rest of the planet in darkness. Many facts force the hypothesis of Asia as the original home of the race, but no remains of the early handiwork have been found there.

A conservative estimate of the oldest remains of our own species is 25,000 years, though some authorities would double and others would treble this estimate. But other species of the *genus* homo have left a few bones which go back very much farther yet. *Paleolithic* (Old Stone Age) men lived in France, Spain, and other parts of western Europe and the gradual advance in their technique of working the rough stones has been represented in the accepted division into periods of which the following six are practically universally recognized, with various subdivisions not so generally agreed upon. The Chellean, Acheulian, Mousterian, Solutrean, Aurignacian, and Magdalenian, are named from the places in France where the deposits were found, sometimes in the gravel beds or "drift" and at other times in caves. It is possible to assert a definite advance or evolution from the first of these through the series, but some of the changes may be due to the sudden incursion of a stranger folk. Indeed, the Mousterian and the two preceding deposits are generally assumed to be the work of another species than ours. The Magdalenian flints, however, were left by the Crô-Magnon people, whose bones have been recovered in sufficient numbers to warrant the statement that they were perhaps physically superior to any existing race of men, being taller in stature and with a larger brain capacity than any modern race. Their mural paintings executed two hundred and fifty centuries ago may still be seen and are the wonder and admiration of all who know them. But whence these people came into France and Spain and why and how they disappeared—guess who will, for there are no facts. (See map on p. 204.)

Following the men of the Old Stone Age came the *Neolithic* (New Stone Age) people, who polished and ground the edges of their axes, knew of fields and grain, erected houses and built huge stone structures, which still remain to puzzle us and pique our curiosity. But it is not clear, indeed it seems a bit unlikely, that the Neolithic men were the same tribes as the men of the Old Stone Age and the setting forth of the separate stages of progress from rough to polished stone may be, after all, merely the record of the different migrations into western Europe, and no more proves or even describes evolution and progress than the description of the culture of newly-arrived immigrants into America proves that we are rapidly becoming illiterate.

It is possible to describe, after the Neolithic, a *Cyprolithic* stage or age, when copper was worked like stone, just as to-day the Andaman Islanders work iron cold as they do their shells. And then the bronze age is reached where the addition of tin hardened the metal till it was a good tool, so good that it was in use clear down to the Homeric age; so good that it was not used or arrived at by most of the peoples in the west or in the islands. With bronze the curtain of civilization is rung up, but the story of the origin of these improvements is yet to be told, if ever it can be told.

Whether the flint workers of France ever went back to Egypt or had any connection with their original home, we know not, but we do know that by the time a city arose in the Nile valley the human glacier had been covering North and South America for six thousand years, the mongoloid and negroid races had not only covered the other continents and islands, had not only been separated long enough to be differentiated, but had also mingled their blood till the problem of the complete classification of the races of the earth is one of the most difficult in modern science. There is no generally accepted classification which includes all the families of men.

OLDER THEORIES OF SOCIAL EVOLUTION

The theory of a gradual and continuous evolution assumes progress upward, due to inherent forces in a people living alone. Another and competing theory insists that isolated homogenous peoples tend to become stagnant and fixed in their organization and that the key to change is to be sought in social contacts due to whatever cause, especially such as migration, invasion or other forms of interaction.

At the present time the conception is a controversial one, and the difficulties met in the effort to make the formulation appealing are very real and very stubborn. Those who oppose the orthodox view of progress or evolution are engaged in trying to substitute objectivity for evaluation. It is interesting to recall, in this connection, that the idea of progress is a modern one. It has been so widely held in our time that it comes to many as a surprise when they learn how recently it came into its formulation. Let us glance at a history of the idea.

Preliterate peoples, having a social organization handed down traditionally by oral transmission, were not concerned with the relation of the present to the past. Indeed, in the sense in which we formulate the picture of our past in order to account for the facts discovered, they did not have a past at all. Preliterate peoples have no history. Mythology is lightly held, is largely art, is thought of in a way quite different from that in which we regard historical epochs. As for the future, they concern themselves with it almost not at all.

When ancient civilizations wrote their chronicles of whatever nature, a momentous change occurred, for literature means contacts bridging time, preserving exactly the words of the dead, and overcoming space. And so we meet early in the history of independent reflection the attempt to answer the question of the sort of path which had been traversed by the race of men.

The first of these that shall concern us here is that of the *Greeks* who formulated the conception of human life as passing through a series of recurrent cycles conceived of in terms of millenniums. What was, had been before, and would be again. Life was thought of as a vast pattern with a repetend. The first age was the Golden Age, then came the Silver, and other baser and still baser elements till the final degeneration should come when the whole process should start over again. The complete cycle was fixed in terms of 72,000 years, at the end of which period it would all start over again. This is, therefore, a sort of anti-progress, a philosophy of degeneration, the whole political, moral and physical world gradually running down like a clock. Readers of Plato will recall his stages of political degeneration, timocracy, oligarchy, democracy, and despotism. To the Greeks progress was unthinkable, and change undesirable.

Quite different was the conception of the course of history when the regal monotheism of *Christianity*, with its doctrine of providence and what Santayana calls the "Christian epic," came into being. To them life was a sort of drama with the scenes all written out and the final outcome known from the beginning. The time-span was shortened to a few centuries; the world had been created by a fiat and was to endure to the Day of Wrath, and after that the curtain would descend and the action be transferred to other stages. And in the meantime, there were no accidents in the providence of God, but there was no progress or evolution in the modern sense.

When the doctrine of evolution began to win its way against the conception of medieval theology, the emotional values which had been furnished by confidence in the essentially beneficent power were abundantly supplied in the attitude of confidence in the moral character of the process itself. Henry Drummond and Tennyson gave utterance to the new-found faith that although the evils of the world are many, they are overcome by manifold forces of good and in the distance there is,

> One far-off divine event,
> To which the whole creation moves.

Science has no quarrel with this formulation, for it is not a scientific question. Evolution as a philosophy has all the values that any philosophy has and no more. As a detailed statement of the origin of anatomical structure, evolution is a scientific hypothesis, and this has been successfully applied to problems in botany, zoölogy, geology and astronomy. When applied to social and ethical problems, it has never been possible to find a method of demonstration, and the facile generalizations of Herbert Spencer have one by one broken down under the increased strain of accumulated facts. Evolution as a philosophy is clearly a child of the wishes, but a child which can only be born to a society whose comfort and prosperity are obvious and undeniable.

The stages through which society has everywhere passed, formulated again and again and correlated with economic organization, familial schemes, moral concepts, religious views and practices, all these have been regretfully abandoned under the strain of accumulated facts which have revealed exceptions, anomalies and *lacunœ* too serious for the theory to incorporate.

But progress is still a good word. Every man knows what it is in reference to his own life and his own purposes. Every society knows what it is to form plans, to work toward them, and to witness their satisfying realization. But progress as the specific achievement of a definite aim is one thing while progress as a steady and progressive

realization of the common good or happiness is quite
another. And in the last thousand years conflict and struggle, warfare and victory, have been so continuously the
experience of human society that it is not difficult to see
that progress must always be stated in terms of the victor
in the contest. It is, therefore, a subjective category.
Optimism is the faith of the successful who believes he will
continue to succeed and that the victories he has won over
his enemies are but the assurances that future enemies will
also be destroyed. It is not too much, therefore, to say
that the older doctrine of progress is losing its attraction
for those who think in terms of the human race. There is
another conception of progress which the scientific age is
formulating which brings the process within the human
will, the human reason, and the human muscles, namely,
*the doctrine of the conscious progress of plans which men
may make, of dreams which the dreamers may dream, and
which by careful and progressively clever methods may be
realized.* From this point of view progress is no longer
the cosmic process realizing itself as all the Hegelians
conceived it, but rather collective purposes, collectively
planned, collectively striven for and, therefore, believed in.
It is a retail and particular process and not a wholesale
and general one. It is the process by means of which we
control our own destinies and analyze our own problems,
making our own plans and bringing them to pass where we
can, in spite of the niggardliness of a step-motherly nature.

The Older View of Social Evolution Tested by Facts

As a doctrine of progress it began in the seventeenth
century; as the doctrine of social evolution it is of the nineteenth century and is the analogue of the anatomical evolution of the biologists applied here not to individual organisms, but to the growth of societies.

The orthodox theory of social evolution is a corollary

of the theory of psychological evolution. As the body can be traced from the very simple forms to a climax in the relatively large brain of man, so mental capacity was assumed to consist of separate states, the lower ones being those occupied by primitive man. Aided by the concept of vestiges, men like Herbert Spencer were able to construct a symmetrical picture with the lower races at one end of the scale, intermediate forms following after, the climax occurring in the geniuses who are the glory of our race.

Presuppositions or the Older View of Cultural Stages

Primitive man, said the representatives of the older view, not only existed in the Old Stone Age, but he also exists today in Australia, in Patagonia, in Greenland, and similar regions of low culture. Culture being the product of the adaptation of the individual to his environment, it was thought high or low as this adaptation was made by a higher or lower order of mind. In the development several stages were clearly distinguished, some formulations of which have become classic and are indeed the intellectual heritage of our literary tradition. A familiar series is the division of cultures into hunting, pastoral, agricultural, commercial and manufacturing. As the facts began to accumulate, subdivisions of these were made and transition stages admitted, but the general framework was not questioned.

An even more familiar designation still current is that which gives the series as savagery, barbarism and civilization. These again are divided by some writers into upper and lower savagery, upper and lower barbarism, and early and later civilization. Again some found it necessary to further subdivide the material, making three divisions of each, lower, middle and upper savagery, and so forth.

The common assumption of schemes of this type is that

culture and social organization result from an interaction between the mind of man, which is assumed to be uniform and constant for a given situation, and the environment, which varies with the climate and physical situation, but which is a definite fixed entity to be "adapted to." The attempt to assign the different peoples to their appropriate place in the scale was repeatedly made with a certain measure of success, the differentia being in each case the possession of a certain specific element of material culture; for example, a bow and arrow or pottery combined with the economic organization or the degree of social integration. It was assumed that the human being who must drink will need a vessel to drink from and that when his mind has developed sufficiently he will know how to adapt himself to an environment which will make him bring into proper relation the three elements of clay, water and fuel. Brought together in proper spatial and temporal relations, clay, water and fuel will produce a pot. The lowest races had no pots because their minds were inferior. When through the gradual evolving power of the intellect they rose high enough in the mental scale, the pottery adaptation took place and they advanced to the higher stage of social evolution.

Analogous assumptions were made concerning the bow and arrow. The bow and arrow is almost unique among human inventions. It has been called the most difficult and most important single material invention. It enormously extended the zone of danger and efficiency of the hunter, gave him a greatly enlarged food supply, and contributed enormously to his feeling of self-confidence and power. But this is not the chief reason for the high place which the invention holds in the minds of the ethnologists. The remarkable aspect is that it is impossible to see through what stages the invention has passed. With a spear it is different. A poor spear is still a spear. A poor pot has some value as a pot. But a poor bow and arrow is practically worthless. Now, the origin of the bow and arrow is unknown, being prehistoric, but many tribes exist who are ignorant of it. The older theory assumed that elastic wood or similar material, straight shafts and twisted cord were put into their proper relation when the mind of man had advanced far enough in self-direction and mechanical skill to make this possible.

And so on through the series. Domestication of animals is higher than pure hunting and was assumed to have arisen when the scarcity of game and sufficient mental power occurred together. And so with agriculture.

The technical name for this theory is *Independent Origin*. Through the American continent the bow and arrow was used. It is also present everywhere through Africa. It was not assumed that the Africans learned to make the bow from the Americans, or *vice versa;* but rather that peoples in both cultures developed the instrument at their proper stage.

Criticism of the View of Independent Origins of Culture

Such a theory has all the attractiveness of symmetry and simplicity. It held the field for a long time and has by no means been wholly abandoned. Questions, however, began to arise when careful studies revealed certain spatial relationships that suggested difficulties. If a map of North America be drawn with reference solely to the manufacture of pottery, the areas where the art is known are practically continuous. A line drawn from the northern part of Arizona roughly in a northeastern direction will separate the area of pottery south and east of this line from the area of no pottery on the north and west. It might be assumed that the people of the north and west were inferior to the others, but the question was raised very early whether the art of pottery had not been introduced and taught to the remote tribes by some who had learned it or discovered it. The situation is quite similar regarding the bow and arrow. There is a large section of Oceania where this invention is unknown. That part of Oceania where bows and arrows are used is contiguous on the map with Malasia and the continental areas which have possessed this instrument from prehistoric times. Here again the assumption is entirely tenable that the lower races are those who have not yet advanced to the stage of culture where the invention could occur, and it is entirely thinkable that this division of mankind into lower and higher might occur were the given peoples not entirely contiguous in the areas they occupied. In the case of the bow and arrow, however, complications affecting the theory of progress early appeared.

The Andaman Islanders are admittedly among the most primitive of people, having no agriculture nor any pastoral life, living off native pigs, fish and turtles and with the very simplest form of social organization. They have, however, excellent bows and arrows with which they are very skillful. Certain Polynesians, on the other hand, whose social organization is complex and who have chiefs and kings, are ignorant of the bow and arrow. Moreover, the weapon is used in the northern tip of Australia and the Australians have long been considered among the most primitive of peoples.

One more instance may be cited, the discovery of iron. There is still current a scheme of social evolution which gives as the stages stone, copper, bronze and iron, and there is no question of the validity of this division of cultural elements in the case of the inhabitants of western Europe in the prehistoric times. But when we consider that throughout the continent of Africa iron was mined, smelted and forged and that in North America, where there are the richest deposits of iron in the world, no use whatever was made of it, it is impossible to avoid serious questioning concerning the implications of the orthodox theory. The Iroquois Indians or the Pueblos, the Aztecs or the Cherokees, when carefully studied, appear to have no lack of mental ability. Dr. Eastman, a native Sioux Indian, began to learn to read in his adolescent days and fourteen years later was awarded the degree of Doctor of Medicine from Harvard Medical School. It is seriously to be doubted whether the absence of iron-working in America is to be ascribed to a low degree of mental power, and when we look at the map of iron culture it is again a continuous area which appears.

The Diffusion of Cultures

The accumulation of facts of this nature has led to the theory known as *Diffusion*, which would account for the spread of inventions in terms of contact with other peoples. That it is possible to trace the march of an invention in all its meanderings and in the absence of written records no one would assert. But given the appearance of an efficient weapon like the bow, and assuming contacts and migrations so that one group might learn from its neighbors, it would be easily possible to find the bow and arrow introduced to a people of low mentality, but entirely absent from those of superior ability because they had not had the good fortune to be reached by its influence. This whole subject is still a matter of controversy among specialists, but a sufficient number of indubitable connections have been made out to impair seriously the older formulation of the evolution of material culture.

The older conception of the life of primitive peoples has been modified in two important respects. First, there has been apparently a continuous mobility, continued to our own times, which gives a very different picture than that presented to a scholar who wrote fifty years ago. We know of voyages of more than a thousand miles of the South Sea Islanders in their ocean-going canoes. Anthropologists now regard the American Indians as kindred of the Mongolians and assume that the Fuegians on the southern tip of South America are there because of a slow, but unceasing, migration from Alaska throughout the whole length of the two continents. Similar itineraries have been made out of the two wings of the Bantu race, who started somewhere in northeast Africa, divided to the east coast and the west coast and met again in the region of the Cape. Far more recent have been the migrations of the Maoris, the date of whose arrival in New Zealand has been provisionally fixed at the thirteenth century A.D. We think of the modern era as characterized by free movement of peoples, and this is true, but it is merely a question of degree and rate of movement. The prehistoric world is now everywhere pictured to us as characterized by migrating, advancing, intermingling peoples. So thorough has been this process that many anthropologists assert that there are no pure races left on the earth, least of all the Africans.

A second change in our conception is the realization that an element of culture can travel from one tribe to another without the presence of those originating it. The researches of Boas have shown that tales and myths are relayed from language to language and can be traced through thousands of miles, those finally telling them having no familiarity with the language in which the stories first originated. When Stanley came down the Congo River he found food plants that had been domesticated in South America growing thousands of miles inland, hundreds of miles beyond where white men had ever penetrated, having been relayed within the last two hundred years. Another instance of this process is to be found in the journey round the earth of the practice of smoking tobacco, which was brought to Europe in the seventeenth century, spreading soon to Asia, extending to the whole of Africa, finally reaching, by way of Siberia, the Indians in Alaska, who had been ignorant of it. The practice made its spiral circuit of the globe in one century, and this before the advent of steam power.

All this has much to do with the theory of cultural evolution, making it easy to see how many or even most of the elements of the culture of a people may have been borrowed. It is now easy to see why the pygmies are expert archers and some of the Polynesians still spear-throwers, or why Soudanese ex-slaves can read Arabic though Marquesans remained preliterate.

There is another cardinal feature of the classic theory of social evolution that has been fatally criticized in recent years. It is the assumption that with a given economic organization or stage there would always be found a corresponding political, moral and religious stage of ideas and institutions. Much has been made, for instance, of the position of women with reference to the degree of advancement in culture. The most primitive women were assumed to be lowest in status and each advance in cultural development was assumed to be reflected in a higher stage with reference to this particular culture element. But when the facts began to accumulate, this simplicity did not appear. Some primitive tribes do, indeed, treat their women with scant consideration, beat them imprison them, and make them into beasts of burden. But these are not always the lowest tribes. Indeed, they are never the lowest tribes. The simpler peoples are the kindlier. It is among the more advanced that harshness becomes striking. The writer has seen the wife of an African chief sitting on the ground wearing a punishment fork on her neck made from a heavy log whose continued weight could be nothing short of torture. But these people were agricultural with half a dozen breeds of domesticated animals and a high degree of skill in metal working, weaving and wood carving, while among the Iroquois Indians, who were in the Polished Stone Age, the matrons of the tribe had great freedom, much dignity, and a high degree of political and administrative responsibility and power. It is unnecessary to multiply instances of this sort for the statement is unquestioned that the economic, the social and the religious development do not run *pari passu.*

Conclusion

What shall we say then? Has there been no evolution or development of social life and organization? Is it not possible to see any progress in the march of the human race? It does not follow because the older explanation of evolution is unsatisfactory that no continuity or improvement can be made out. The psychology of invention is not easy to write. In fact, it is perhaps forever impossible to formulate it, for invention is something new and to hit upon something new and original always has the quality of the accidental, by which we mean the not understood. No one knows who invented the art of working iron. It is certain that it was not a white man and it is not impossible that two or more men could have done it independently and in far-separated regions. Whoever it was, he passed it on to others until now it has become the foundation of our modern civilization. The invention of iron, however, does not seem to be any measure either of mental capacity or of high culture. The Eskimo, who had neither metals nor stone, who understood neither weaving nor pottery, stands conspicuous among primitive peoples as an industrious, efficient, and highly moral person. For many millenniums—the guesses at their number are very wild for the only measure is the very small scale map of geology—primitive man over all the earth lived on a level

of culture which, with all its variations, can hardly be separated with any degree of scientific confidence into higher
and lower.

THE FACTS OF SOCIAL EVOLUTION

We return, then, to the assumption which underlay the
formulation of our fathers in the age of faith, namely, that
the human race is approximately uniform in mental en-
dowment, and that progress and change are not to be correlated with or explained by any assumptions of increasing
mental capacity. This was the assumption made by those
who formulated the course of history in terms of divine
providence. As we have already seen, throughout medieval
thought and surviving still in evangelical circles it was assumed that the whole course of human life, creation, fall,
redemption, the last day, and the millennium, were all
conceived in the mind of God, who had made of one blood
all the nations of the earth. The modern doctrine of
progress is but a translation of these terms into the scientific language of the nineteenth century. The span of the
years was enormously lengthened, and the details of the
scheme were loosened noticeably, but the steady growth,
and irresistible improvement of social and moral ideas was
steadfastly believed in and still has many able and earnest
advocates. They no longer speak of a millennium to be
inaugurated by the visible, literal, bodily *Paroursia*. It is
nevertheless confidently believed by many that a goal of
change exists.

The emotional value of such a conception is unquestion-
able. Moreover, it accorded so well with the doctrine of
biological evolution that scientific warrant for this emotional faith was easy to procure. The doctrine of the survival of the fittest when applied to social phenomena is
very full of comfort for it is always preached by those
who have survived, and who thus assert their fitness by a

scientific indirection. Ethnocentrism, the tendency to make one's own culture the measure of all others, seems to be everywhere present, but to those who are enabled to contemplate all peoples from the standpoint of the whole the relativity of these measures soon appears. For progress is relative to the ideals and wishes of a nation or a race, and in a world as bloody as ours progress has meant the death and destruction of many from whose point of view there has been of course no progress. To a Bulgarian Christian the history of the Balkans in the last two centuries shows more progress than appears to his Turkish neighbor. The Cherokee chiefs fighting a rear guard action against encroachment and injustice can only believe in progress by taking the point of view of their enemies. The dog who is running for his dinner, and who is gaining on the rabbit running for his life is indeed making progress, but the rabbit, soon to be a victim, could he think, might not so define it.

Considerations such as these have done much to discredit the earlier generalizations. Particular progress in specific activities is obvious, but whether in general, and on the whole this could be asserted must depend on the point of view. Those of European culture, including Americans, whose ships dominate the seven seas and whose flags float over even the barren wastes of the poles are hardly justified in identifying their own achieved ambitions with the fate of the race as a whole. Moreover, warning voices in no small number have been raised, calling attention to the new types of degradation in our slums, new forms of slavery in factory and brothel, new types of discontent, and what is more serious if true, the physical degeneration of the race in a land where the most miserable one-sixth produces more than half of the children. It is necessary to reëxamine older formulations if we are to escape the fallacy of our own prejudices.

PRELITERATE AND LITERATE CULTURE

Leaving out of account the new terms of value, what descriptive changes can be described? One of the most significant is held to be the *invention of alphabetical writing*. Diverse as they are, preliterate peoples are much more nearly alike and have far more in common with each other than with those who, possessing written literature, we call civilized. Now literacy is an institution arising out of certain inventions, originating in very circumscribed spots, and spreadng over the earth. The missionaries of our own generation have abundantly proved that all peoples can learn letters. The unclothed Bantus can be seen sitting in the shade of their forests reading a book which six months before had seemed to them like unintelligible magic. Literacy then is not the result of capacity, but a tradition handed on from one race to another and from fathers to sons, which can be traced back to the dawn of history. Whatever social changes literacy represents, it has this much of the fortuitous. That literacy does make profound and fundamental changes is increasingly evident, for the written word remains. In the records of the past the fathers speak, and the writing or scriptures are venerated among all peoples. Such facts mean that a continuity of tradition, an enlargement of consciousness is possible to such a degree that it amounts to a difference in kind. A race without letters has no history, merely tales and traditions. A people without history is like a man without memory who lives from moment to moment. It is possible, therefore, to make one grand division in the evolution of man at the period where writing begins. For the preliterates, different as they are from each other, may all be characterized by certain common traits. They live in small groups; they are relatively isolated; they are uncritical of their own culture, and their lives are lived in a social atmosphere where magic and superstition have reached the saturating point.

The Effect of the Invention of Writing

When writing appears several things happen. The past lives on in the inscribed leaf. Isolation both in time and space begin to give way. Contacts multiply and cities begin, and with the growth of cities, and the complexity which necessarily results a new dimension is added to human life. W. J. Perry and Eliott Smith have brought forward many facts in support of the notion that city life with its coöperation and consequent accumulation of savings, or capital, furnishes the culture medium in which were evolved both slavery and war. This theory is too recent for a final judgment to be passed upon it, but the complexity of large aggregations with its division into classes, and the stratification which finds its ultimate expression in Europe in feudalism, and in Asia in the caste system seems quite undeniable. Now one aspect of social evolution which we can attribute to writing is the systematizing and fixing of the moral, spiritual and social ideas and customs. Preliterate societies are erroneously assumed to be fixed and immovable. Properly understood, the opposite statement is more nearly true. All tribes have food taboos, religious and ceremonial practices, but none are so fixed or have endured so long as those of peoples who have written them down, for writing fixes the old and the old always tends to become sacred. Every one who is familiar with preliterate culture recognizes the helplessness of their traditions in competition with an organized and systematized competing system, whether the missionary be Christian, Mohammedan, or Buddhist. The preliterate villager has no effective defense against his confident assertion of a fixed and hoary tradition.

It is not meant to assert that writing necessarily represents a higher stage of culture. Indeed the point has been repeatedly made that writing introduced too early may be a great bar to progress. The elaborate and meticulous ritual of the toilet would probably have been abandoned long ago in India, had it not been written down in the sacred books. The irrational dietary laws of the Hebrews which forbade ham, but permitted grasshoppers, survives to our own day only because of the literacy of their fathers.

If one term must be chosen to characterize the effect of writing, it would be the tendency to *absolutism*. Preliterate peoples live in a world of magic and environing spiritual beings, but these are evanescent and shifting in their existence. The introduction of writing mitigates in no sense the tyranny of superstition, but it does erect it into a system giving stability and permanence and the prestige of former generations. In the enthusiasm for classic culture, it was for a long time customary to deny superstition and magic to the Greece of Pericles, but careful researches forced the admission that this praise is undeserved. The life of medieval Europe is so well known that it is impossible to minimize the place of magic and superstition in their culture. Indeed it is easier to list the likenesses between the civilization of Egypt, Greece, Rome, and Medieval Europe than it is the differences. Differences there are in plenty, but through all the variety there appears the common characteristics, expanding political units, stratified society, sacred books wherein the superstitions of their fathers are glorified and a uniform and pathetic dependence upon unknown supernatural beings and influences, gods, devils, gnomes, fairies under the control of witches, necromancers, shamans and priests. If the word were not already preëmpted by historians for another meaning, it would be convenient to designate all that period of human history from the earliest civilization in Egypt to the sixteenth Christian century as the medieval period, for middle period it certainly is, intermediate between the unorganized and half-conscious life of the preliterates, and the modern age, the keynote of which is control.

The Stage of Control of Natural Forces

John Dewey somewhere remarks that the idea which appeared in Western Europe in the period we know as the Renaissance was the most important invention of the human mind save perhaps the invention of language itself. This idea he says was the conception that the forces of nature can be used and controlled to satisfy and increase the wants of man. Our third period of social evolution is then the *age of control* or the *modern period*. It is unnecessary to admit that anticipations of this may be found as far back as paleolithic man, and the art products of Egypt and of Europe would have been impossible without a measure of this spirit. But despite these facts it remains true that the whole center of gravity of our culture has shifted in the last three or four centuries from dependence and submission to conscious invention and control. Wissler characterizes European culture by three terms: universal suffrage, education, and invention, and these are all different manifestations of the modern spirit which is homocentric, self-reliant, and when true to itself devoid of superstition. The history of this transition has been but recently written, nay, it is still in the writing. One of the most important documents of this history is White's "History of the Warfare Between Science and Theology in Christendom," a history which he could not completely write since he is dead, and the warfare not yet finished. The different chapters in this book are accounts of the several battles in that war, and may all be brought under our formula. Astronomy began as an effort to reduce to a mechanical formula the movements of those terrible points of light whose influence on our fathers is still reflected in the names of our week-days. The comets were transformed from portents of wrath to harmless streams of luminous gas and insane people have by the touch of science been transformed into hospital patients who formerly were the helpless hosts of disembodied demons.

It is difficult to exaggerate the magnitude of this change. The control of astronomy, the control of navigation, the control of agriculture, these are all such commonplace assumptions of our culture that we need at times to be reminded of the prescientific methods which secured the safety of ships by prayers and offerings, and the fertility of the soil by magical and erotic ceremonies. The dawning conception that human nature itself is the result of social interaction, and that psychology and sociology can become natural sciences opens up the hope that by taking thought human nature itself can be controlled. War, poverty and crime which were formerly defended, apologized for and even conceived as a part of the divine plan, appear to our modern eyes as problems to be solved, as challenges to the technique of control which scientific men persistently seek.

That this is another and higher stage of social evolution can admit of no doubt. The modern scientific world is a different world from the medieval universe where all the evils were a part of some higher plan of extra-human powers. The medieval has this in common with the preliterate world, that the emphasis of importance is always on some other life—superhuman or infrahuman, but never human. The preliterate world is social, mythological, magical; the medieval mind added theology and metaphysics, the modern conception is positive and scientific.

The True Conception of Social Evolution

The conception of social evolution is then that of a dependence upon new inventions and discoveries, but these are not necessarily or chiefly material. Human brotherhood is as much an invention as a steam engine and democracy as real a discovery as electricity. The history of these concepts bears a direct relation to the growth of social organization and Shailer Mathews, who speaks of theology as transcendentalized politics, has shown in the clearest way the relation of our concepts of the universe to our social life. The preliterate world is still in many parts a godless world for gods cannot exist where there are no kings. When civilization appeared the pantheon of each culture reflected the political structure of its people. In this our modern age the revolution has been so recent and so fundamental, that concepts and imagery for the religious symbols of a democratic people have not appeared in any satisfying form. The transmitted scriptures of our fathers give us a fossil vocabulary of a medieval world, a vocabulary which fits but poorly the needs of our day. The new wine of democratic ideals is endangered in the old skins of medieval vocabularies.

And yet nothing would be more erroneous than the assertion that social evolution has outstripped religion or that science is taking the place of religion. The dreams of our dreamers are as splendid as any Syrian prophet's inspiration. The faith of a modern advocate of peace on earth, or a modern prophet of social reform is of the same quality as that which has made the record of the Hebrew prophets perennial fountains of courage and hope. There is this difference, however: the preliterate faith was a dumbly despairing trust in capricious and precarious spirits; the medieval faith was a humble and contrite surrender to an arbitrary and powerful external deity; the modern faith is a trust equally sublime and of the same quality, but having for its instrument a scientific technique to be expected and sought for. The love of a man for the life of his child is of the same quality in all three cases. But in sickness the Melanesian relied on pure magic; the early Christian on fervent prayer; the modern man on preventive medicine and its attendant sciences.

And so we conclude that there has been evolution and progress after all, but the formula is *the more or less rapid spread from single centers of diffusion of particular inventions and discoveries in the material and spiritual worlds.* Some of these inventions have been evil, and some of the change has been regressive. The great discoveries have always presented new problems, some of which have not yet found solutions. The doctrine of progress here presented would view our present evils and future perils in the nature of a challenge to our inventive genius and associated creative intelligence. With the conviction that this is not a stage play which is already rehearsed with the final consummation already certain, but a real fight with the issue in doubt, and a real struggle into which the high-hearted can throw themselves with all the devotion of the ancient heroes, we work earnestly to find a technique which will enable us to achieve the object of our faith, the bringing in of a better world.

EVOLUTION OF THE MORES

We come finally to the problem of the evolution of culture narrowly defined, which is the distinct subject-matter of sociology. The key to the understanding of this question lies in a study of custom, and the stages through which it passes. Now the customs of a group are its habits, analogous to the habits of a man, and arising out of the normal tendency to repeat an act in the same way time after time. Custom has all the advantages of individual habits. Attention is economized and efficiency results, for energy is more effective if a channel already exists. Custom likewise has the disadvantage of habit for habits are hard to break and not all habits are good.

The stages of custom are now generally agreed upon.

They begin as folkways, the unconscious uniformities of behavior which arise in every society. The folkways are in the beginning never formulated, and are only partially attended to for not only is the tendency of habit to become unconscious through long repetition, but also the beginning of habit may be entirely unconscious.

The second stage of development of the folkways is known among sociologists as the *mores,* which while still unformulated are more conscious and always in some degree emotional for the violation or threatened violation causes concern or resentment. The folkways, which are mere usages, exist in all societies alongside the *mores,* which are all but universal, but not quite so. It is possible to find isolated societies on small islands like the Andamans where hardly any of the folkways have risen to the conscious and emotional level of *mores.* This means that resentment at the violation of the folkways has not occurred because the violation has not sufficiently often taken place. There is no penalty for murder among the Andamans, that is, no set penalty. If the murderer be a man of influence he sometimes withdraws himself for a time from the camp followed by some of his friends and stays away until the matter blows over, after which the whole thing is forgotten. Thus even the *mores* seem to require a certain degree of interpenetration of groups to bring the folkways to the conscious level of morals.

The third stage of development is a double one for it takes two directions, one individual and one social. On the individual side group morality passes into individual morality; custom becomes conscience. And here again our formula seems adequate. Conscience among completely isolated peoples is so rare as to be negligible, for conscience is an appeal which the individual makes from the group to his own ideas, setting himself in opposition to others or feeling guilty because of his refusal to obey the voice of

his people. The literature of isolated preliterates seems
to warrant the assertion that a homogeneous group hardly
ever has the problem of dealing with one who criticizes the
customs of his people or refuses to fall in with their wishes.
Modern civilized life with its company of martyrs, heroes,
rebels and independent thinkers has obscured the obvious
principle that individuality presupposes a sort of dual
membership or at least a dual influence. Conscience is not
merely the voice of the group in the soul of a man; it consists in the warring voices of two groups, or of multiple
social influences contending in a single breast for allegiance and supremacy. In modern life this is not difficult
to see, particularly if we take into account the influence of
literature, for the reading of books is, as already remarked,
a kind of conversation with the past or at least with the
absent.

The social aspect of the third stage of development is
the passage from the unformulated mores to the organized
institutions. Now an institution as Sumner points out
involves a concept and a structure, the concept being the
abstract symbol, the product of reflective thinking, and
the structure being an organization into a formulated and
systematized arrangement of personnel and material. This
process may be illustrated in religion, which begins in the
unconsciously formed folkways arising out of the quest for
food, the defense against enemies, and the crises in life and
in nature, such as birth, death, winter and storms. The
folkways thus gradually crystallized are called *mores* when
they come into consciousness and are rationalized. Likewise, the phenomenon of conscientiousness in religion is
most easily observed when two religions simultaneously
solicit the allegiance of one man or when a strange custom invites him to disregard or criticize the religious practices of his fathers. And finally, while among preliterates
there are no religious institutions, yet among moderns there

are no religions without institutions, our conception of religion being bound up with our ideas concerning the church, the mosque, the synagogue or the temple.

There is one important modification of the statement that these phases are stages of evolution. It is not the whole truth to say that isolated peoples are governed by folkways, conflicting preliterate groups by *mores,* and modern peoples by institutions. The folkways are as much a part of modern life as of the most backward people, and the *mores* exist even where institutions are most numerous. The *mores* do not replace folkways, but are superadded to them and institutions do not replace the *mores,* but exist alongside both the earlier forms of control.

Consider a book of etiquette describing the social usages of the most refined society. Such a book might be defined as a set of written directions enabling members of a lower stratum to behave consciously as the members of a higher group behave unconsciously. The manners of the superior social group are folkways and are absorbed without effort by the children, being enforced by no severer penalties than lifted eyebrows of pained surprise or gently smiling approval. For every society develops its unintentional customs, which, if they continue long enough, may pass over into the stage when they are expected so confidently that they are enforced by severe penalties although the penalties may not become exact and formal. And when the customs reach another stage they may pass into legal enactments, thus reaching the institutional stage. The prohibition law may be thought of as the efforts of part of the nation to impose their *mores* upon the whole. In many cities it is now illegal to alter the direction of a motor car without extending the hand horizontally. The custom having proved desirable, it became a law and passed quickly into an institutional phase.

These three stages may indeed be thought of as being

preceded by another, a sort of instinctive morality whereby as in a figure nature punishes violations even though society is organized in favor of them. The polyandry of the Todas, the infanticide of the Solomon Islanders, and the birth control of the Bobangi are rapidly causing their extinction. But neglecting this phase, the four stages of evolution may be set down as folkways, *mores*, conscience, institutions followed by the disorganization and breaking up of these latter, and the reorganization into new systems. But all this is obviously concerned with the form only, and not at all with the content, and it is the content of morality which is important. If we inquire whether there is a development or evolution of *mores* on the side of content the matter becomes very difficult. No practice which we deprecate or abhor has been without moral approval among some people, somewhere, at some time. The Greeks thought it highly moral to kill sickly children; the Fuegians kill their aged parents as a sacred duty; and the Australian offers his wife to his guest less he be considered inhospitable. To this day head-hunters in many communities feel ashamed until they have raided a sleeping village and decapitated the helpless victims. The content of the *mores* depends upon the fortuitous constellation of forces, economic, political and social. A sudden change in circumstances will make a good practice immoral. The exploitation of the children in our factories is but one of many examples. Alexander Hamilton is quoted as praising the new machine age because it brought the opportunity for gainful employment to all the people, especially those of tender years.

Moreover, folkways and *mores* are as much the object of import and export as are material goods. Witness our Australian ballot, our German Christmas tree, our Chinese game of Mah Jongg. The development of the content of customs is never a simple evolution, but includes the sudden acquisition by one folk of what has been slowly built up by another. If the tribes of earth had each remained quite separated from all the rest it might be possible to have described their cultural evolution with more confidence, but the mobility of races and cultures is the outstanding phenomenon. Whenever for any cause the mobility is decreased customs tend to harden and become stable. This not only characterizes China and India but to some extent Medieval Europe, and perhaps the most salient feature of the cultural life of our time is the rapidly increasing tempo of alteration.

The development of the folkways may be into *mores* and thence into institutions, but the change does not necessarily take this direction, for the manners of people may change or disappear or undergo substitution while still remaining on the level of mere usages. Likewise the *mores* undergo constant modification, decay, intensification, or substitution without necessarily ever becoming institutionalized. And as for institutions, they are always being altered, and some of the changes are very slow. Moreover, some institutions pass back into the life of a people, as mere customs, remaining sometimes as vestiges, surviving in a few instances in the games of our children before they disappear entirely from human life.

And when there are sudden and dramatic changes in institutions these are never the result of immediate causes alone, but may be thought of as a sudden eruption due to long continued and increasing pressure, or as a tree long decaying may be overturned by a sudden gust of wind. They can only be comprehended if we consider that the slow process of undermining has been going on for years. Revolutions have occurred in all ages of history from Egypt to Russia, and the formula seems everywhere to apply. The revolution may mean moral advance or it may not. The judgment in each case depends upon the judge. But the revolution is the breaking up of an old organization

and the tendency of human society is to reorganize itself
as best it may, and as soon as it can.

Every institution, like every organization, involves the
expression of some attitude, and the suppression of others.
The equilibrium obtained is never permanent for the temperamental equipment of the rising generation is never
identical with the adults who are in command, and the
temperament of new leaders introduces at times a disturbing factor. Moreover, widespread communication gives
increasing opportunity for new and disturbing changes,
and these always make for disorganization.

Modern life is perhaps most truly characterized as involving an increasing rate of change whose *tempo* is speeding up in a geometrical ratio. More changes have taken
place in the last generation than in the previous century,
and greater changes in the last hundred years than in the
preceding thousand. Whether this be good or evil depends upon the outcome, and concerning the outcome no
one may dogmatize, for possibilities of growth involve possibilities of decay, and men who may continue to advance
beyond middle life are also subject to the perils of disorganization in a far greater degree than their ancestors.

Ten years ago it would have been possible to secure widespread agreement to the proposition that the increased communication of our day has led us to an era of democracy,
and it is clearly true that our units are larger, and the
area of our sympathies includes more people than ever
before. ''Men have always believed that it was right to love
your neighbor as yourself—the difficulty has been to agree
on who your neighbor is.'' There is dawning an age of
humanity. Our circle of brotherhood sometimes includes
the planet and there is some warrant for saying that there
has been a continuous expansion of the family sympathy.
Nations were once everywhere considered to be above all
moral law, but now we plan a Parliament of Man. It may

be that institutions will culminate in a Master Institution, with justice and liberty, equality and democracy as cosmic ideals.

And yet it would be very easy to fall into an error, even here. In 1917 the hearts of men were lifted up because they saw visions of universal democracy following the last of all the wars. To-day we are a sadder race. The tragic discords of Versailles and the bitter hatreds that have arisen out of our intense reactionary nationalisms have already produced a flood of articles, pamphlets and books glorifying isolation, defending the exploitation of the weak, and repudiating democracy. The future even of democracy cannot be foretold out of hand. If and when we work out an adequate social science, we shall be able to predict, and to control because we shall know the processes and the mechanisms. Until then the issue is veiled. The outcome depends on the vision of our seers and the skill of our leaders, as well as upon the inscrutable movements of the cosmic processes whose outcome, being inaccessible to our knowledge, remains the goal of our faith.

CHAPTER X

MIND AND EVOLUTION

By Charles H. Judd [1]

IT is not the purpose of this essay to enter into the discussion of the philosophical questions which center around such words as intelligence, mind and personality. The metaphysical nature of intelligence is a subject on which there have always been and always will be fundamental disagreements among men of different temperaments and different creeds. Science is content to treat intelligence as it treats other forms of reality. It aims to understand its various manifestations and to discover, if possible, the results which issue from its operations. Science asks such questions as these: What are the various levels of intelligence; what are the marks of higher and lower minds; what effects do minds produce in the world; what are the conditions under which these effects are most readily produced, what the conditions under which minds are impeded in their action? Science in confining itself to such questions adopts here the same attitude that the chemist adopts when he studies the behavior of oxygen and leaves to the philosopher the problem of the ultimate nature of matter.

Approaching the study of intelligence in this strictly scientific spirit, we note first of all that there are many different grades of intelligence in the world. We observe the lower animals and note that they show by their behavior that they do not think and plan as do human beings.

[1] Director of the School of Education and Chairman of the Department of Psychology, the University of Chicago.

There are, however, other respects in which the animals are in no wise inferior to us. They have very keen organs of sense, in some cases their senses are superior to ours. Putting all these facts together we conclude that thinking and planning involve something higher than the mere reception of impressions through the senses. The comparative method aids us in this way in defining the nature of intelligence.

Another fact which is observed is that some animals seem to profit greatly by experience while others show only meager powers of retention of experience. Evidently the ability to hold what one has had in consciousness is an important asset, characteristic of the higher animals and of man. Indeed, some students of animal life have come to the conclusion that the most significant fact in animal evolution is this fact that memory or the power of retaining experience and of reinstating experience in consciousness raises animals at some stage of the organic series out of the lower levels of existence and sets them on the path that leads upward to self-control and constructive thinking.

Comparisons between intelligences of different levels has sometimes been resented by persons who are afraid that human life will lose something of its glory and exclusiveness by being compared with animal life. It is the steadfast belief of many that in the world of mind if not in the world of bodily structures human nature rises to an absolutely separate station in the universe.

It is not difficult to find in the nature of consciousness itself an explanation of the attitude referred to in the foregoing paragraph. One's consciousness is in an important sense of the word an exclusive inner world. Shut up in our individual worlds of thought and emotion, each one of us looks out upon a universe which has for us only secondary value as contrasted with our own inner beings. One's own personality includes those likes and dislikes which one cannot fully express to another person; it includes those fleeting ideas which come and go and make up the current of individual mental life. One's personality is made up of that world of memories and ambitions which no other human being can share directly, a world so vivid and real that philosophers of all ages have described consciousness as the primary reality.

If a human individual looking out from his conscious experience tries to understand the relation of his personality to the rest of the world, he finds that the range of his vision and consequently of his understanding is circumscribed by barriers which seem to be impassable. Let any one ask himself what is the origin of his personality and it becomes at once evident that the inquirer's present experience does not reach far enough back in the history of the world to comprehend its own origins. Indeed, the range of personal comprehension of one's self is even more limited; one does not remember in adult life the experiences through which one passed in maturing from the lower forms of infant and childish consciousness to the higher forms of thought and imagination characteristic of later life.

Shut in by the insuperable barriers which surround personality, the human mind has initiated studies which are intended to answer the questions which this inner world does not answer for itself. Just as the sciences which deal with material things have reached out beyond the range of direct observation and have reconstructed our ideas by bringing together and coördinating facts which no eye can see, so psychological science has attempted to give a view of the nature of personality which no individual mind can directly supply to itself. Geography, for example, tells us that the earth is a sphere revolving through space. We accept this description though no eye can see the round ball or feel its motion. Again we know that matter is made up of atoms, not because we see these minute particles but because scientific methods of extending direct observation have carried us beyond the ranges of possible vision or touch. Something of the same type is accomplished by psychology which brings to each individual a broader view of his place in the world than he can attain by looking directly into his personal experiences.

The history of humanity shows that there has been great reluctance on the part of human minds in accepting scientific conclusions about the nature of the mind itself. We are all so self-centered, so much in contact with ourselves, so intimately acquainted with our own emotions and thoughts, that any scientific assertions which connect our personalities with less vividly experienced realities seem to us to be in some sense degrading. The world has learned to accept the pronouncements of science about the shape of the earth and about everything else, and has grown complacent in the adoption of statements which can never be verified by direct personal observation. Not so, however, when it comes to explanations of ourselves.

The difficulty which thus stands in the way of a science of consciousness is one of the major reasons why there is at the present time a violent recoil against the doctrine of evolution. People do not want to be classified by science with animals or things. They know that they live in an inner world which is unique and they are prone to feel that science must not invade this world with assertions which cannot be verified by personal experience. Science, on the other hand, is quite sure that it is possible to classify personalities, to show that some are of low grade and some of high; to show that intelligence contributes in various ways to progress and to the conquest of the world. Science is quite sure that many of the conditions which bring about the improvement or disruption of personalities can be described and controlled. In short, science is making steady progress in dealing with mental nature just as it has in the course of time learned to deal with atoms, with living cells and with celestial bodies.

PHYSIOLOGICAL PSYCHOLOGY

The first requirement when one is going to deal scientifically with a series of problems is to find a method of scientific inquiry. The science of psychology found a productive method of procedure some seventy-five years ago by studying a series of very tangible facts which throughout the animal kingdom and throughout the life of individual human beings stand in close and constant relation to conscious processes. A special name was coined for the branch of science which was thus developed; it is the name "physiological psychology." Physiological psychology notes that there are certain anatomical and physiological facts which always parallel intelligence. It notes the fact that a child born with a defective brain will have an abnormal personality. It notes further that if a brain becomes diseased, there will be abnormalities in the inner world of consciousness. If the blood stops flowing to the brain, as in fainting, consciousness lapses.

The methods of physiological psychology have nowhere been more productive than in the comparative studies of mental life. The lower animals have been found to possess brains which are relatively simple as compared with the brains of the higher animals. Indeed, it has been found that a certain portion of the brain shows in its structural variations a very direct relation to the scale of intelligence. This part of the brain is called the cerebrum.

The characteristic fact about the cerebrum is that it does not receive its impressions directly from the organs of sense nor does it send its orders for action directly to the muscles of the body. The cerebrum is what may be called an indirect organ. It is the part of the nervous system which stands over and above all the lower organs. The lower parts of the nervous system are in direct contact with the eye and the ear and the hands and tongue. The cerebrum gets its incoming stimulations from the lower centers. The human cerebrum is a vast network of nervous tissues. It unites and organizes the manifold currents sent up by the lower centers and after combining all the incoming currents, sends them out to the lower nerve centers to be distributed to the organs of action. The cerebrum can be compared to the central office of a huge manufacturing concern. To this central office go up reports from all parts of the plant, and from the central office come well-digested plans for the guidance of future operations.

If the method of study employed by physiological psychology is applied comparatively to the different levels of animal life, it is found that while animals are fairly well equipped with the lower nervous centers, that is, direct sensory and motor centers, they are supplied with very meager cerebrums. The frog, for example, has eyes and uses them very efficiently for the direct business of life which is catching insects for food, but the frog has a very small cerebrum. This is parallel with the fact that the frog has very few higher mental processes. A frog does not think over the problems of life. It has no part of its body where this thinking process can be housed. It has no cerebrum in which impressions can be worked over into ideas and judgments.

As we pass up the animal scale from the frog to man, the parallelism between the power to carry on complex mental processes and the extent of the tissue in the cerebrum bcomes one of the most clearly established of scientific facts. The dog is higher than the sheep; the gorilla is higher than the dog. These facts are at once facts of cerebral anatomy and of comparative intelligence. The cerebrum not only relates the various currents of stimulation which it receives at any given time from various parts of the body, but it is a vast storehouse of past impressions and past combinations. The animal which has a large cerebrum is less dependent on the impression of the present moment than is the animal with little cerebral tissue. Here again we find it to be a fundamental fact of animal life that there must be tissue for every function. The animal that has little brain tissue can perform only lower functions. The animal with more tissue and a higher physical organization has the bodily structures which are essential to those higher forms of life which we name consciousness and intelligence.

The revelations of physiological psychology help the individual to understand his place in the world. The human being who inherits from his ancestors a large cerebrum, complex in its structure, has inherited possibilities of a high and complex personality. This personality consists not in the direct reception of impressions from the organs of sense, but rather in the power to remember and to think. Personality is made up of the power to compare, to relate, to organize impressions and memories into new combinations.

EXPERIMENTAL PSYCHOLOGY

There is a second scientific method, no less productive for the understanding of personality than the method of physiological psychology. It is the method of experimental study of behavior. In 1878 Wundt, the great German psychologist, set up the first laboratory for the experimental study of mental processes. Since that time the world has cultivated very widely experimental methods in this field and has made great progress in the solution of many practical problems of controlling the development of mental processes.

Perhaps the most striking examples of the successful applications of the experimental methods can be found in the scientific study of the processes of learning. Suppose that the scientist puts two animals or two human beings under conditions that are alike and observes the rate at which they learn how to react. For example, if food is put behind a door that will open only when a certain secret latch is touched, the scientist can observe two animals with perfect assurance that the one which is more intelligent will succeed in learning how to open the door with fewer trials than the animal which is less intelligent.

It may be remarked in passing that the method of comparison which science has thus elaborated is no different in essence from the ordinary method of estimating intelligence which society universally employs. The business world esteems as able and intelligent the man who learns quickly and retains the results of his experience. All that science has done is to refine and elaborate an ordinary method of classifying personalities.

Scientific studies have been made of the learning processes of all kinds of animals from the lowest unicellular forms of life to the most intelligent human beings. The methods of learning of each have been recorded in what are known as learning curves. In general it may be said that the typical learning curve among the lower animals is one which rises gradually through many trials and errors to the point where such skill of response as the animal is capable of exhibiting is attained and there the learning process stops. On the other hand, a frequently observed fact in the highest forms of learning such as are exhibited by human beings is a sudden rise in the level of efficiency due to the fact that the individual develops an idea and thus succeeds in guiding his behavior through a superior form of intelligent control.

The study of learning processes has proved to be highly useful in directing education. It is found that the immature stages of learning in human beings are in many cases of the trial and error type. It has been possible to describe with exactness in certain systems of learning clearly distinguishable stages which exhibit very different rates of progress and very different types of learning. Thus in learning to read, children pass through various stages and the intelligent management of their education requires a discriminating recognition of these different stages.

A striking illustration of the difference between ordinary experience and scientific knowledge regarding the nature of intelligence is supplied by such cases as that referred to in the foregoing paragraphs. Many persons have learned to read but do not know what processes they have passed through. Individual mental life is made up of a series of absorbing efforts. One very seldom sees one's own development as a complete series. The series cannot be recognized by the immature individual because so long as there are unattained levels of maturity, he cannot foresee them since they lie beyond the range of his experience. The mature individual, on the other hand, does not recall with any degree of exactness or detail the stages through which he has passed. The scientific knowledge which is now worked out about the total process of learning is therefore a body of material transcending the experience even of those who have passed through the process.

Comparative studies in which like methods have been applied to animals and men show that the most striking fact about intelligence is that it has been steadily increasing with the evolution of the animal series. Furthermore, the laws of intellectual progress seem to be continuous, coupling human life with life of the subhuman level.

The History of Human Intelligence

We come, however, at a certain point in our comparative studies to a unique fact. With the appearance of human intelligence, there seems to be a sudden acceleration in the evolutionary process. Human minds, while they have many traits in common with the minds of animals, show such superiority and such unique characteristics that many students of the problem of evolution accept the conclusion that we are here in the presence of an absolute break.

In general, science does not hold that human intelligence is a fact outside the evolutionary series. Science holds, rather, that in the progress of the animal series a stage was reached at which conditions were favorable to a consummation of the tendencies which had been slowly in the making further down the scale. The lower animals have some memory, some powers of communication, some powers of thought. The animals have cerebrums where the higher processes of nervous organization can go on to some extent. At the same time that the lower animals exhibit these structures and the abilities which enter into what we call intelligence, they are dependent in the main for their adaptation to the world about them on their teeth and claws. In other words, animal life is in the main one of direct adaptation. But direct adaptation is relatively slow. It takes generations to evolve powers of fast running or to evolve muscles serving to make a powerful bite. On the other hand, indirect adjustments resulting from thoughts and intelligent planning are instantaneous in producing efficient adaptations. The animal which can react to its environment intelligently can forego strong muscles and rapid flight. Wit and cunning are surer reliances than brute force. The evolution of intelligence is therefore sure to bring a new epoch in the life of the world. Once the higher forms of intelligence appeared, the process of evolution takes an entirely new turn.

Science has not always admitted the importance or the newness of the turn. There has been a tendency in some quarters not only to say that the laws of learning are alike in animals and man but to say that they are exactly alike. There has been a tendency to find in animal life the analogies by which human society is to be explained. Where the tendency has manifested itself to treat human evolution as nothing more than a direct continuation of animal evolution, it is probably true that too great devotion has been exhibited to the principle of continuity. Human intelligence is higher than brute intelligence and, because it is higher, it has struck out on new lines. It has by its very effectiveness changed the balance of life. We no longer fit ourselves to our environments by the slow process of evolving new organs; we adapt ourselves to our environments by more intelligent use of those organs which we have.

How Inventions Guide Human Progress

The science of psychology in studying human minds requires, therefore, a method which is supplementary to the methods of physiological and experimental psychology. The supplementary method which has proved to be most productive of illuminating findings is the historical or social method of studying the progress made by the human race under the guidance of intelligence.

In order to connect our discussion at this point with the comparative studies which were reviewed in earlier sections of this essay and at the same time to mark off human evolution from that which preceded it, let us take up briefly the evolution of the use of tools. Animals are typically destitute of the intelligence necessary to the invention and use of tools. The few instances in which monkeys are reported as throwing missiles and elephants are reported to have used levers do not materially affect the generalization stated above. Animals do not use tools. They accomplish whatever they do by the use of their claws and teeth. Theirs is the method of direct attack.

The consciousness of animals is always centered about the one object toward which attack is directed. They are at all times in the same mental condition as the human being who is reaching out for something which he is about to grasp. Their consciousness has only one focus. All this is in sharp contrast with the conscious experience of man who has learned to use tools and to think in terms of tools.

The man who first invented, or, rather, discovered a tool, had two foci in his thinking, and it was because of his superior powers of thought that he could include in a single experience both an object of attack and a tool or the means of attack. Or, putting the matter in another way, a tool is the product of a double view.

Consider the case of the first sharp stone which was used as a knife. Many an animal had cut itself on sharp stones and had growled at the experience. Doubtless the most primitive men took at first scarcely more than the brute's attitude toward sharp stones. But the day came when the first discoverer of the tool, a human being with a big cerebrum, stopped and considered the sharp stone which had cut him with something more than mere anger at his own suffering. He had enough brains to see the stone in new relations. If it cut him, he could take it in his hand and use it to cut the things which he wanted to attack. There are two foci of consciousness in this being's mind, the thing to be attacked and the means of attack.

With arrival at this stage, the process of evolution has taken a most significant turn. From this point on the animal kingdom will not evolve merely by putting longer and sharper claws on the extremities of animals in order to fit them better for the struggle for existence. From this point on, evolution will be determined chiefly by the powers of the cerebrum which is able to put together two centers of attention. It is enormously more advantageous to the individual and to the race that tools should be developed than that claws and teeth should be perfected.

The process of evolution which began with the first discovery of a tool has gone steadily forward. The second stage was that in which man, having discovered the utility of a sharp stone, began to improve his implement by shaping

STAR CLOUD IN SCUTUM SOBIESKI.

This shows one of the richer parts of the Milky Way. No two stars visibly separated on this photograph can be actually nearer to each other than millions of miles.

THE SPIRAL NEBULA IN ANDROMEDA.

This is probably one of the nearest of the spiral nebulæ, being the largest in angular diameter, but its distance is reckoned in thousands of light years. It is approaching the earth with a speed of 186 miles per second. This photograph was taken by G. W. Ritchey with the 24-inch reflector of the Yerkes Observatory.

it so that it fitted his hand better and also became in its cutting edge a more vicious instrument for attack on his enemies.

The period when men were learning artificially to improve their tools was one freighted with momentous consequences for personality. The ferocious primitive hunter with his brutelike interest in personal attack began to change into the thoughtful, attentive observer of materials. He soon began to cultivate a sense of utility and proportion. He became observant of the hardness of various kinds of stones. He learned to watch and imitate his competitors and his fellow workers. He began to think of the devices of trade by which he and the hunter might coöperate to mutual advantage. This led to the development of new forms of cunning, the cunning of the artisan and the trader. It must have led also to a new form of social interest, the interest in coöperative living.

Primitive manufacturing next led to the search for raw materials. The quarryman and ultimately the prospector for metals became a specialized member of the rapidly differentiating community. The mental traits and ambitions of these new workers in the social group were very different from those of the hunter and of the manufacturer. The consequences to personal experience of the division of labor began to appear.

There is an important sense in which division of labor is both a consequence of human thought and a powerful influence in molding human experience. Division of labor results naturally from the fact that even though men have cerebrums large enough to include in a single act of attention more than one idea, they are not capable of indefinite extension of thought processes. As soon as a human being gets absorbed in one kind of thinking, he tends to turn all his mental energy into that one channel. The manufacturer becomes so absorbed in fashioning materials that he is no longer interested in the chase. The prospector is so bent on the adventure of discovery that he is unwilling to settle down to the labor of the artisan. Human beings tend to specialize, hence the division of labor.

As soon as a system of coöperative life is evolved because people have specialized, it becomes a dominant influence and determines what future generations must do in order to survive. Division of labor is to-day more than a result of the tendency to specialize; it is a compelling scheme of life. It has turned upon the very human nature that brought it into being and dominates to-day our education, our industry and our very personalities.

LANGUAGE AND NUMBERS

The historical method of studying mental evolution can be further illustrated by following another line of thought. Human minds active in securing a more complete adaptation of the individual to his environment have invented certain ways of thinking which after being invented have come to be the most important factors of the environment in which individual personalities grow up. Language is such an invention; the number system is such an invention; so is money. For our purposes it may be well to follow briefly one of the more recent of these inventions where we know quite definitely the steps which have been taken. We might study language as the most important of the inventions of this class, but the facts regarding language are lost in great measure in remote antiquity. We shall choose the more recent evolution which is exhibited in the number system.

Number was invented by men when the desire to hold in mind a large collection of objects led to the substitution of a tally for the image of the thing itself. The shepherd who wanted to know whether all his sheep were in the fold could recall each member of the flock by name, or he could mark off a tally for each sheep and let the simpler tallies be his guaranty that the full flock was there. In tallying man first used the digits which were the most accessible, simple objects against which to match the objects that he was interested in recording. Everywhere in the world the decimal number system bears testimony to the fact that man early counted with the aid of his ten fingers.

Later, man perfected his number system and added devices of combination and recombination. He made himself an abacus and he devised a system of written tallies. All these devices were relatively primitive and did not permit thinking to go very far in the direction of a cultivation of systems of calculation. If one wishes to realize how primitive were the methods of calculation even among the Romans, let one try to set down a problem in multiplication in these numerals. For example, let one try to multiply LXXVII by XIX. This case makes it very clear that higher forms of thinking require the perfection as well as the invention of devices for the support of thinking.

In the particular case under discussion, the perfected device came into Europe from the Orient some time during the latter part of the medieval period. We still give recognition to the source of our number system by the use of the name Arabic numerals. The virtues of this system need no discussion here. The system is a brilliant illustration of the fact that human intellectual progress has been the result of inventions and refinement which are wholly different in type from the inventions of material tools. Here is a purely intellectual instrument which has become through the exercise of human ingenuity just as truly an element of our environments as is any material object. In fact, we force children to fit their thinking to the Arabic numeral system somewhat more vigorously than we force them to deal with any of the material things about them. We find that ability to calculate accurately and fluently is so desirable an acquisition for all persons who are to live in a coöperative society that we not only preserve the Arabic numerals but we also give them the highest social sanction. It has come to pass that our modern world could not exist if it did not have as one of its instruments the Arabic numeral system. Our modern science depends for its formulas and its methods of calculation on this system of number. So also do our commercial system and our industrial system.

The evolution of the special form of intelligence which we call the number system could be paralleled by other examples no less striking. Money and the still more abstract system of credit which underlie our whole commercial life are at once products of human ingenuity and dominant facts in the control of every modern human being.

Machine industry is another similar example. We are a machine-working generation. It is said of us that we lack enthusiasm for our individual tasks because the processes of standardization have gone so far that the worker has lost his sense of ownership and responsibility. The worker goes about his task shiftlessly and without those satisfactions which came to the handworkers of an earlier age. Yet while we thus deplore the evil effects of machine industry on the individual, we recognize the fact that machine industry is one of the ripest products of inventive intelligence.

Abnormal Psychology

There are other paradoxes in the evolution of intelligence. Human cerebral structures have grown so complex that they not infrequently break down and exhibit abnormalities which can be properly described as the immediate consequences of their own elevation to high levels. The lower animals exhibit some abnormality of nervous response, but it is only the highest forms of nervous organization which can fall into the disorders which are exhibited by human personalities.

The study of intelligence must of necessity therefore be accompanied by a study of mental pathology. An abnormal mental state can be thought of as one in which certain conscious elements which under other conditions would be constituents of normal states are distorted in their emphases or in their relations and as a result produce states which are unbalanced. Thus a person who gets the idea that he is very rich or very powerful will in ordinary life check his belief in these ideas by the facts about him, and will, as a consequence, not commit himself to the acceptance of the suggested ideas. On the other hand, the abnormal individual may accept the suggested ideas and act upon them because his abnormal condition prevents him from checking his belief. Thus abnormal and pathological forms of mental life by their distortions often throw new light on normal processes.

There are a great number of mental states in which the perfectly normal individual must be described as temporarily lacking in those relations among the elements of his mental life which are necessary to the exhibition of what we call intelligence. We are all temporarily lacking in rational balance when we are dreaming. Dreams are found to be very closely related to many forms of mental pathology. In dream life the emphases are controlled and biased by considerations which in normal life are fully or largely subordinated. Sex impulses are perhaps more than any others predominant in this as well as in certain spheres of mental abnormality. The use of analytical methods which have grown out of the study of dreams has brought relief to many cases of incipient functional disorder. While there is ground for the statement that the emphasis which these analytical methods have laid on the sex impulse is exaggerated, there can be no doubt that human ills have been alleviated as a result of the careful analysis of all of the border-line phenomena related to mental pathology.

EMOTIONS

One general field of human consciousness which has been found to be of great significance alike to the student of pathology and to the student of mental life is that which is designated by the terms feelings and emotions. Until recently psychology dealt only in vague terms with the emotions. It had no adequate understanding of their nature. It was one of the most notable services to science of that great American psychologist, William James, that he supplied an explanation of the emotions. His discussion was published at about the same date as that of a European physiologist, Lange, and the doctrine on which they agree usually bears both their names. Modern psychology accepts in its essential outlines the James-Lange explanation of the nature of the emotions.

Adopting the formula of James for the purpose of contrasting as sharply as possible the scientific explanation of the emotions with the popular explanation, we may say that a person's sorrow is conditioned by the fact that he weeps; it is not true that one is first profoundly distressed and afterwards weeps. Or if we express the matter in slightly different terms, we may say that the common man believes that the order of psychological explanation is as follows: I receive an impression; I am sorry and weep. The true order as known to science is this: I receive the impression and because of the upheaval in my motor system, I weep. As a direct result of the upheaval in my motor system, there arises in my consciousness an emotional state of sorrow.

How did science come to the conclusion that motor disturbances are primary and not secondary? Lange, the physiologist, noted that there are frequent cases in every sanitarium where the melancholia of the patient is traceable not to any mental cause whatsoever but rather to a disturbance in the activities of the visceral organs. James noted cases in which impressions produced flight before fear set in, thus proving that the reaction is primary and the emotion secondary, rather than the reverse. The laboratory with its experiments has amply supported the theory of the emotions worked out on the grounds mentioned.

The importance for society of this fuller understanding of emotions is unlimited. We now know that people cannot work skillfully when they are emotionally excited because emotional excitement means a far-reaching motor disturbance. Put in terms of practical management of industry, this means that people cannot work well when they are angry or afraid. Long years ago the Federal Commissioner of Labor, Carrol D. Wright, commenting on the strikes in the coal region, said they were purely psychological in their causation. He was trying to explain an important economic disturbance. The psychologist coming at the matter from the other side reaches a similar conclusion. Wherever action is involved there will always appear a corresponding mental state. If action is disturbed and complications arise through the appearance of conflicting reactions to objects other than those which should engage attention, then the emotional state will be unfavorable.

Many a laborer is inefficient because he is afraid. He is thinking about the possibility of losing his job or about the dangers of old age and his fear is physiologically a state of muscular tension which interferes with his digestive activities or with the movements of his hands so that his skill is diminished. Fear is one of the most general experiences of human life. The menace to socity of machine industry is in large part due to the physical condition of tension which it induces. Furthermore, the fears of modern life are difficult for the individual to master because they arise in no small measure from the fact that human nature is in a relatively new and artificial environment. Human nature was more at home in the primitive conditions of the open. There if one began to be afraid, one ran away. Fear was useful, it stimulated an appropriate reaction. Under the artificial conditions of modern industrial life, if one is afraid one may want to run but one cannot do so and even if one could, one would not protect one's self in this way. Fears grow under the artificial conditions of life just because they are ineffective.

It would not be just to drop the example which we have been discussing without calling attention to the fact that a scientific understanding of the nature of emotions has led to the intelligent treatment of many cases of abnormality and to the betterment of emotional conditions among normal individuals as well. Once society realizes that fear is a fact of behavior, preventative measures can be devised and have been adopted in plans which provide systematic recreation and promote the development of ability to relax.

LIMITATIONS OF PSYCHOLOGY

There is one negative statement which must be made in order to make clear the province of the science of mind. Psychology has no method of determining where or how consciousness began. The history of thought affords many examples of speculations on this matter. Some students of mental phenomena have been led by what they deemed to be the logic of their discoveries to believe that consciousness is coextensive with life. Some speculative thinkers have gone even further and asserted that consciousness is a part of the being of every particle of matter. On these speculations psychology as a science has no part.

Nor can the science of mind determine the possibilities of the existence of consciousness after life. If there is a continuation of the processes of thought in a human being after his cerebrum ceases to function, science cannot make a report on the process. The methods of psychology are those which have been illustrated in the foregoing paragraphs.

Within the range of phenomena with which science can deal there can be no doubt that the general process of evolution is constantly exhibited. The particular kind of evolution which goes on in the world of mind is defined by science on the basis of its observations and experiments. There is nothing in the abstract doctrine of evolution which can guide science; the facts must be sought in detail and must be arranged into a coherent system of explanatory principles.

SUMMARY

It has been the purpose of this paper to suggest the general outline of such a coherent system. This purpose has not been achieved unless at least two general conclusions have been made clear. The first is that everywhere mind shows that series of progressive stages which justifies the assertion that there has been from the beginning, and is now going forward, a steady process of evolution. The phenomena of mental life always fall into an orderly series of evolutionary progress. It matters not whether one studies the broad series of phenomena of animal behavior or the narrower series of facts which go to make up the life of an individual animal or human being. Everywhere mind is moving through a series, either upward to more and more complex adaptations or downward through processes of dissociation to lower and often abnormal levels. It requires no long historical series of manifestations of intelligence to establish the principle of evolution and yet when the long series is examined it is found to obey the general law. Our first conclusion, then, is that in the large and in the small, where mind is, there is evolutionary change.

The second conclusion of this paper is less likely to command general assent. It is laid down, however, with great assurance in view of such facts as have been presented. It is as follows: With the appearance of the highest form of intelligence, that is, intelligence of the type exhibited in human beings, evolution took a new turn. The method of animal adaptation which is the pattern throughout the lower ranges of organic life gave place to a method of adaptation in which intelligence is dominant and physical development of special organs of adaptation receded.

If this second conclusion is accepted, it becomes evident at once that the application of the doctrine of evolution in the form commonly designated biological evolution cannot be accepted as valid. Biological evolution derives its statement of the details of the processes which it reports from the study of animal life. The notions which biological evolution formulates about coöperative communities, for example, are the notions derived from the studies of beehives and ants' nests and from the study of flocks of birds or herds of buffaloes. Biological evolution cannot properly carry over its statements and apply them to human society. Human society has language and number systems and tools and money. The methods of coöperative living among men are wholly different from the methods of coöperative living among bees or buffaloes. To attempt to explain all the facts of life by a single formula is to fail to recognize the fact that evolution is not a single process but a name for the general fact of serial change. The particular facts of evolution in any given sphere are the facts which science must discover and formulate before it attempts to tell what evolution in that sphere means. Social theory must therefore rise above biological evolution. The grounds for this assertion have been presented in numerous examples cited in this paper.

It may not be amiss to point out that the foregoing is in no sense of the word a denial of the validity of the general doctrine of biological evolution. The plea for a scientific statement of human evolution, distinct from the doctrine of evolution as applied to lower forms, does not cut off one word from the doctrine of biological evolution in its own sphere. Nor does it put man outside the sphere of biological evolution. Man has certain characteristic traits which do not depend on the action of his cerebrum. All that is asserted here is that the complete account of the higher forms of life, especially of man, will have to be based on a study of these higher forms and their performances, and this complete study will find that intelligence is the determining factor in man's life.

PART II

SCIENTIFIC COÖPERATION WITH NATURE

CHAPTER XI

RECENT CONTRIBUTIONS OF MEDICINE TO HUMAN WELFARE

By JOHN M. DODSON [1]

IN the knowledge of the causes of disease, of their modes of transmission from one person to another, and in the ability to control the spread of disease, medicine has advanced more in the last sixty years than in all previous time. This is a surprising statement when one reviews the long line of distinguished physicians from Hippocrates to Pasteur, a period of twenty-three centuries. Its explanation is largely to be found in, first, the epoch-making demonstration by Louis Pasteur, a French chemist and scientist in the early sixties of the last century, of the germ origin of certain diseases; second, in the remarkable development of physics, chemistry and the biological sciences in the last half century; and third, in the enunciation by Virchow, a German physician, in 1851, of the cellular hypothesis, that is, that the cell is the unit of all living beings, the tissues and organs of the body being made up primarily of cells.

THE GERM THEORY OF DISEASE

That disease in larger animals might be due to their invasion by microörganisms, was a theory which had been entertained by physicians in almost every generation but it re-

---

[1] Executive Secretary, Bureau of Health and Public Instruction, American Medical Association.

mained for Pasteur to make positive demonstration of this fact. How convincing and irrefutable was the proof he offered, first in the case of a disease of the silkworm; later in a disease of sheep, called charbon; and still later other experimenters in a long list of diseases, is illustrated by the phenomena observed in diphtheria, the causative germ of which was discovered by Klebs and Loeffler some twenty years after Pasteur's first work.

From the throat of a person suffering from this disease, a minute speck of matter is taken with a sterile swab, or platinum wire, best from the false membrane which usually forms in the back part of the throat in this disease. If this is now introduced into some clear, sterile beef broth, with a small admixture of blood, in a test tube, the tube being then sealed in such a way as to prevent all possibility of any germs gaining entrance to the broth from the air, or if the swab be smeared over the surface of a similar broth made solid by the addition of gelatin or agar-agar, and to which a small amount of blood has been added, there will appear, if the test tube or gelatin plate be kept for a few hours at about the temperature of the human body, a cloudiness in the broth, or a grayish film on the gelatin plate. On examination of a droplet of the broth under the microscope, one sees myriads of minute rodlike bodies, which assume a characteristic appearance when stained with one of the aniline dyes, like methyl blue or violet. These are the bacilli which cause diphtheria.

If now from this first culture, a minute droplet be taken and placed in a second test tube of broth, the same cloudiness appears in a few hours. From the second tube a third is inoculated, from this a fourth tube, and so on until the experiment has been carried to say the fortieth or fiftieth generation. In this will be found myriads of the same bacilli. And now, if a portion of this fiftieth culture be smeared on the throat of a susceptible animal or person, invariably there follows, in a few hours, redness, swelling, usually the appearance of a false membrane, and the animal has difficulty in swallowing, fever, and the other symptoms which were present in the patient with diphtheria from whose throat the first culture was inoculated. A speck of matter from this animal's throat is found to contain the same bacilli, and these can be grown in culture media in the same way as were those from the sick person; the final culture smeared on the throat of a third susceptible animal or person will again produce the disease.

This crucial experiment has been repeated in the case of diphtheria and of other diseases, many thousands of times. It is proof that diphtheria is due to the invasion of the body by diphtheria bacilli, which is as conclusive and unassailable as any experimental evidence in the whole range of human experience.

DISCOVERIES OF ROBERT KOCH

By a chain of evidence similar to the above, Robert Koch, in 1876, while he was a district physician in Wollstein, Germany, demonstrated the bacilli of anthrax; in 1878 the fact that wound infections are due to germs; and in 1882 that the bacillus of tuberculosis is the cause of consumption. In connection with these discoveries, he devised a method of isolating germs in gelatin plate cultures, outside of the body, and of differentiating them by stains which made possible, accurate, conclusive demonstration of the causal relationship of other germs to several diseases. He also enunciated the so-called postulates of Koch, that is, the several steps which must be taken to prove that a particular disease is due to the invasion of the body by a certain microörganism:—first, the organism must always be found in the body of a person suffering from the disease; second, it must be cultivated outside the body through several generations, so as to separate it from all other substances which may have been present in the material first taken from the sick person; third, the disease must be produced in the body of an animal, human or otherwise, which is susceptible to the disease by the injection of a culture of the germs; and finally, the germs must be found in the body of the animal thus inoculated as the result of the disease so produced.

It has not been possible to establish every one of these four postulates in the case of every disease in which a causative germ has been quite certainly discovered, because to some of these diseases only man is susceptible, and for other reasons. Other methods have since been found, however, which confirm such a causal relationship, conspicuously the effect on such germs of the blood serum of persons who have suffered an attack of the particular disease in question.

Following the discovery of the bacillus of tuberculosis, there came in the following year, 1883, the discovery by Koch of the germ of cholera, and, within a few years those of typhoid fever, diphtheria, gonorrhea, pneumonia, tetanus, or lockjaw, the plague, leprosy, malaria, syphilis and many other diseases, until at the present time, some twenty-eight diseases have been proven to be due to bacteria; six are due to protozoa, the lowest forms of animal life; several others are due to germs which are so small that they pass through the pores of a porcelain filter. At least ten other diseases are quite certainly due to microörganisms but the nature of these has not been definitely established.

Toxins and Antitoxins

That germs produce their evil effects in most diseases by the generation of poisonous substances—so-called toxins— was first proven in 1888-89 by Roux and Yersin in the case of diphtheria. Soon thereafter it was discovered by

Behring and Kitasato that the tissues of the body of one who is afflicted with this disease respond to the action of the diphtheria toxin, by producing a substance which antagonizes or neutralizes the toxin and so defends the body from the invasions of the disease.

This antitoxin may be produced in a susceptible animal, like the horse, by injecting the animal with repeated doses of the diphtheria toxin, of gradually increasing strength. Most important of all, this antitoxin is produced in large excess and, being in solution in the blood, can be obtained for use in the treatment of human beings, by withdrawing some of the blood from the animal, allowing it to clot, and then injecting the pale, straw-colored liquid, which separates from the clot—the serum—into the body of a person, arresting the disease when it is used early, or preventing its appearance in one who has been exposed to it. When the antitoxin is injected to forestall and prevent diphtheria it is said to confer an immunity to the disease.

IMMUNITY

The question of immunity is one of the most interesting and important chapters in the literature of medicine. The theories by which it is sought to explain infection and immunity, chief of which is the so-called side-chain theory of Ehrlich, are somewhat complicated and need not here be discussed. It is sufficient to say that many human beings are, or may be made, immune to the effects of certain of the germs of disease, as shown by the fact that when fully exposed to contagion, or even actually inoculated with such germs, no disturbance follows.

To several of these diseases a considerable proportion of the human race are born insusceptible, for example to diphtheria, or to scarlet fever. This is called a *natural* immunity. In many instances immunity may be *acquired*—either by an attack of the disease, contracted in the usual way, or produced by the inoculation of the person with the germs of disease, usually with their virulence artificially lessened. Any one of these methods results in arousing the cells of the body to produce the substances which antagonize the germs of the disease or their toxins. The protection thus acquired is known as *active immunity*. When, on the other hand, as in the case of diphtheria, there is injected into the body the blood serum of an animal whose blood already contains an excess of antitoxin there results a *passive* immunity. The length of time during which one is rendered insusceptible to a disease by such acquired immunity varies widely in different diseases and in different individuals. One of the recent important advances in the treatment of diphtheria is the use of a carefully balanced mixture of diphtheria toxin and antitoxin which induces a passive immunity of immediate effectiveness and an active immunity of longer duration.

The hopes which were aroused by the discovery of the antitoxin for diphtheria, that corresponding remedies might be found in other diseases, have not been realized in any large measure. Only in epidemic cerebrospinal meningitis and to a much lesser degree in the plague, tetanus and possibly one or two other diseases, have curative serums been discovered. In these and several other diseases, however, some degree of protection against persons who are exposed to infection, as in times of an epidemic, can be secured by such inoculations.

TRANSMISSION OF COMMUNICABLE DISEASE

Not less important than the discovery of the germ origin of the diseases which are communicable from one person to another, is the knowledge which has been gained of the means by which they are transmitted. That diseases may be conveyed by insects has been abundantly established. In some of these, insect transmission, so far as we now know, is the only way, as in malaria and yellow fever, each of which is conveyed by a particular genus of mosquito. The knowledge of this fact, together with a study of the breeding habits of these insects, has made it possible, by destroying their breeding places, and preventing their access to infected persons on the one hand, and to well persons on the other, to exterminate yellow fever almost completely, and to bring malaria largely under control. Vast regions of the earth, especially in the tropics, have thus been made habitable for the Caucasian.

Other insects, among them the fly, the louse, the tick, the flea, and the bedbug have been shown to be frequent offenders in the transmission of disease, and disease has been greatly lessened by making use of this knowledge, as by the wholesale delousing of civilians and soldiers in epidemics of typhus fever during the World War, the exclusion of flies from human beings and their excreta, and in other ways.

DISEASE CARRIERS

Another discovery of far-reaching influence is that persons who have apparently completely recovered from a communicable disease, or even those who seem never to have suffered therefrom, may harbor the germs, in the secretions of the nose and throat, as for example, in diphtheria, or meningitis, or in the discharges from bowel or bladder as in typhoid fever or cholera. The detection and segregation of such carriers and their treatment so as to rid them of infectiousness have done a great deal to minimize the spread of infection.

MODERN SURGERY

The fact that wound infection, which made the surgery of former years so disastrous in its results, was due to the invasion of these wounds by germs, was inferred by Mr.

(afterward Sir) Joseph Lister from the studies of Pasteur in fermentation and putrefaction, a few years before Koch proved this relationship. To him we owe the institution of antiseptic (later changed to aseptic) surgery, which added so enormously to the possibilities of surgery. Previous to the work of Lister in the early seventies, surgical operations were mainly confined to the extremities and to the superficial parts of the body. The opening of the body cavities, the head, chest, abdomen and the joints was almost invariably followed by infection, suppuration, often gangrene and very frequently by general blood-poisoning. The introduction of clean, antiseptic surgery, by which such invasion of the surgical wound was prevented, changed all this. Any portion of the body may be invaded by the surgeon with safety, and the most vital organs exposed and operated upon with impunity. What a tremendous advance has been made in surgical procedure because of this, is known to all intelligent people.

The Rapid Growth of the Sciences Fundamental to Medicine

The Science and Practice of Medicine consists in the application of the facts, principles and methods of certain fundamental sciences to the study and treatment of the diseases and accidents of the body. These foundation sciences are physics, chemistry, and a long list of sciences which have to do with living things—the so-called biological sciences. The most fundamental concepts of physics and chemistry have been revolutionized in the last thirty years, and as the processes which constitute life are all physical and chemical in their nature, the biological sciences have also undergone profound changes. Bacteriology, physiological chemistry, pharmacology and several other sciences are almost wholly a creation of the science of recent years.

The discovery by Fischer of the chemical composition of proteid substances, of which albumin is an example, and the newer knowledge of the chemistry of starches, sugars and fats, have made possible a clearer understanding of the processes of digestion, absorption, secretion and of those changes which take place in the innermost cells and tissues which constitute what is called metabolism—that is, the building up and breaking down of the body substances which is constantly going on. From the facts thus gained the science of nutrition has been developed, which enables the physician to determine what foods and how much of each of them should make up the diet of the individual at different ages, in health, or while suffering from one of several diseases in which diet is an important factor.

The so-called vitamins, whose chemical nature is still much of a mystery, have been found to be essential to life, their absence or deficiency leading to such disorders as lack of growth, or to scurvy, or a peculiar inflammation of the nerves known as beri-beri or to other disturbances.

PHARMACOLOGY

Pharmacology, which is the study of the effect of drugs on the living body, has shown that very many of the drugs which constituted the major part of the armamentarium of the physician of a generation or two ago, are wholly without the power to produce the results which were attributed to them. One of the most remarkable contributions of this science to medicine is the famous 606, or salvarsan of Ehrlich. After trying more than 600 preparations of arsenic, he finally hit upon one which will kill the germ of syphilis without seriously damaging the cells of the body. It is an almost certain cure for this widespread disease when used early and in the right way. A few similar remedies have been found by the same experimental methods which are effective in other diseases.

The Ductless Gland

The human body contains many glands which elaborate a fluid secretion which is either a waste product, to be eliminated from the body, or a substance which is of use in the digestion of food, or in some other way. The secretion of such a gland is discharged through a duct, into the digestive tract, onto the skin or elsewhere. It had long been known that there were scattered through the body a number of glandlike structures without ducts—the ductless glands. The only possible outlet for any secretion of such glands is that they should be taken up by the blood and conveyed to other portions of the body.

About the middle of the last century Claude Bernard discovered that the liver, in addition to its function of elaborating bile which is discharged into the small intestine, also had the power to store up carbohydrates in the form of glycogen. At about the same time Addison, in London, discovered a new and remarkable disease attended with the bronzing of the skin which was associated with degeneration of a pair of small ductless glands situated just above the kidney. Many years later it was found that extracts of these suprarenal glands had the remarkable property of contracting blood-vessels and certain muscles when given *in exceedingly small quantities.* Late in the last century came the discovery of the relation of the thyroid gland in the neck to cretinism (a form of congenital idiocy) when the secretion is deficient, and to Graves disease when it is excessive. These discoveries have made it possible to relieve cretinism by feeding thyroid extract, the effect of which on such children is one of the most remarkable phenomena in medicine. Where the activity of the gland is excessive, surgical removal is often productive of relief or cure.

The pituitary body, a peculiar body on the under surface

of the brain, is another ductless gland whose secretions have a profound influence on the rest of the body. The most recent and perhaps the most beneficent of these internal secretions is one elaborated by certain cells of the pancreas and known as *insulin*. That the pancreas is intimately related to the disposal of sugar in the body, and therefore to diabetes, had been proven by the fact that removal of the pancreas in animals as well as certain diseases of that organ is invariably followed by the appearance of an excess of sugar in the blood and its appearance in the urine. That it is not due to the external secretion which is discharged into the intestine through the duct of the pancreas was made evident by the experiment of injecting the pancreatic duct with paraffin, or of ligating it. This leads to the shrinking and ultimate disappearance of the cells which line the ducts and produce the digestive secretion but does not result in the death of the animal or in the appearance of sugar in the urine. It was made evident by these observations that the substance which has to do with the proper combustion and disposal of sugar in the body tissues is not secreted by the cells which line the ducts of the pancreas but by other cells, situated between these ducts, their secretion not being discharged into the ducts but taken up by the blood stream. Dr. Banting, a physician and physiologist of Toronto, Canada, reasoned from these facts that if an extract could be made from these "island cells," (called the cells of Langerhans, because he first described them) and could be so prepared as to permit of the injection of this substance into the body of a person suffering from diabetes, it might increase the power of his tissues to assimilate sugar and so alleviate the symptoms produced when, through failure of this power, certain acid substances are produced whose effects on the body are disastrous. This proved on long and careful experiment on animals to be the case, and this extract of the "island cells," christened "*insulin*," has been made available for use in the treatment of this serious, frequently fatal disease. Its prompt, beneficent effects are quite remarkable. The discovery of *insulin* by prolonged, painstaking and accurate experiment and observation and logical deduction is one of the most noteworthy of recent years.

It is a typical example of the application of the methods of scientific, exact experiment to the problems of medicine, which have come to replace the haphazard, empirical methods of former generations. One of the most fruitful and far-reaching effects of the work of Pasteur is the fact that it made the medical profession no longer satisfied with theorizing and speculation and has actuated workers in medical science with a determination to seek the solution of medical problems—that is, of the nature, causes, and treatment of disease—by the application of the methods of exact science.

The Practical Fruits of Medical Progress

What have been the results of this amazing advance in things medical? Among them are a great diminution in the incidence and mortality of many diseases. Tuberculosis or consumption, the dreaded White Plague, formerly responsible for the death of one-seventh of the human race, has yielded rapidly to the world-wide anti-tuberculosis campaign. In 1900 the annual death rate from this disease was 202 per 100,000; in 1920 it stood at 114. In the United States 8,800 persons were alive on New Year's Day 1921 who would have been in their graves, dead of consumption, had the death rate been equal to that of twenty years previous.

The diseases of infancy and early childhood, but a few years ago the cause of such a frightful mortality, are yielding to the widespread Infant Welfare Movement, and the infant mortality rate in most of the cities of the United States is to-day less than half that of a quarter century ago. Yellow fever, typhus fever and cholera, once frequent and terribly dreaded visitors to our country, have been practically unknown for a generation. Typhoid fever, once the most frequent, dreaded and fatal disease of army life, or wherever large groups of men were gathered, was so controlled during the late war, that it was almost a negligible factor. In all the wars waged since reliable statistics were kept, the losses among the troops by disease in ratio to those lost by the casualties of war, were as 4 or 5 to 1—in the Spanish-American War, as 12 to 1; in the Japanese-Russian War they were reduced by rigid application of measures of modern preventive medicine to the ratio of 1 to 1; while in the World War it was about the same, notwithstanding the great pandemic of influenza which swept the world from 1917 to 1919.

The annual death rate in the United States from all causes was 20 for each thousand of the population in 1900. In 1922 it was only 14 per thousand. This means an annual saving of 600,000 lives in a population of one hundred million. This same figure, 600,000, is to-day the annual death rate from the communicable diseases and it is estimated that if the knowledge we now possess of hygiene, sanitation and the allied sciences could be made effective, this mortality could be at least cut in two.

One of the most important duties and functions of the medical profession at this time is the education of the public in regard to these matters. This its members are seeking to do, individually and also collectively through their county and state medical societies and the national organization, the American Medical Association. Having raised its own standards of education (these requirements have been quadrupled in the last thirty years) it is now undertaking vigorously and by all feasible methods the dissemination of knowledge among the people. The full fruitage of medical progress can be realized only as scientific investigation receives the intelligent, sympathetic, cordial and sustained coöperation of the doctor, the health officer and the public.

Notwithstanding the remarkable and gratifying advance in the medical sciences there is still room and need for great improvement even in the most highly civilized countries. Many more than a million people die in the United States annually and of these it is estimated that six hundred thousand die of communicable diseases, more than half of which are preventable. The lessening of infant mortality does not obtain for the periods before birth, at birth, and during the first month of life; influenza, pneumonia and allied disorders are still unconquered; the ravages of venereal disease have not been materially reduced; cancer, Bright's disease, the diseases of the heart and blood vessels, and some mental and nervous disorders are apparently increasing in prevalence. What are the remedies for this state of affairs?

First, continued, adequate encouragement and support, especially in institutions of higher learning, such as our universities, of the search for new knowledge of disease and its prevention. Second, much more adequate support of public health agencies. Third, diligent cultivation of a more general respect for the rights of others and for law. Fourth, widespread dissemination of existing knowledge of health matters, especially, methods of training and educating the coming generation along health lines in all schools, public and private. It is important that this should be a training not merely toward the dodging of disease but for the purpose of developing sound, vigorous bodies and alert, sane-thinking minds. Such a training makes for individual comfort, happiness and efficiency. The great professions of medicine represented in the American Medical

Association and of teaching as represented in the National Education Association have been engaged for some years, through a joint committee of health problems in education, in seeking ways to better the health conditions and health training in the public schools. If these activities are to flourish, leadership must come, it would seem, first, from the medical profession. The physician must become the family health adviser as well as the family doctor. Second, it must come through organized bodies of enlightened, altruistic persons in such organizations as those devoted to the crusades against tuberculosis, venereal disease, nervous and mental disease and to the promotion of infant and child welfare. For such leadership the world must look mainly to college bred men and women and particularly to those who have consecrated their lives to the service of others in the ministry, in social service, in the field of education, and in other ways.

MEDICAL SCIENCE AND RELIGIOUS FAITH

Have the recent advances in the Medical Sciences developed anything which should disturb religious faith? Not at all.

Many phenomena of nature thought by the ignorant to be due to some supernatural power have been found to be explainable by natural laws, but the conception of a world governed by immutable laws and of a race of men able to discover and interpret them is much more wonderful and inspiring than that of the world where science is unknown and in which the intervention of a special Providence must be constantly invoked.

Medical science—all science—must be constantly seeking to find explanation of the phenomena observed by laws and principles deduced from previous experiment and observation. The instant that science, unable for the time to explain any phenomenon by known laws and principles, invokes the supernatural, that instant it ceases to be science and further progress is impossible.

On the other hand, the first origins of matter, of force and of life, seem as remote as ever from finding explanation in any laws or principles now known to us.

CHAPTER XII

EUGENICS

By Charles B. Davenport [1]

EUGENICS may be defined as the science and art of social advancement by better breeding, or of improving the population by increasing the number of those with valuable racial (hereditary) traits.

Eugenics is to be distinguished from sex hygiene, from hygienic marriage laws, from birth control and other special subjects of propaganda which may, or may not, have eugenical bearings but whose aims are of far more limited scope than those of eugenics in its proper sense.

Eugenics is a branch of biology—of social biology—and its study has been cultivated chiefly by the biologists. The biologist has to look at man in his own way, namely, as an animal, even as the Greeks knew him. Certainly from the standpoint of reproduction and heredity man is only an animal, following precisely those laws which hold for all mammals.

About these laws, especially of heredity, we have learned much in the past two decades. This knowledge has been acquired largely under the stimulus of the rediscovery of Mendel's law of heredity in 1900, when the study of the cell had progressed to a point where the significance of that law was obvious. The new heredity is based on the principle that the determiners of traits are carried in the chromosomes—small bodies in the germ cells—of which one of each kind is received by the nascent individual from

[1] Department of Genetics, Carnegie Institution of Washington.

the egg and one from the sperm. In consequence each kind of chromosome and, ordinarily, each kind of determiner is double in all of the countless cells of which the new body eventually becomes composed. If, in any individual, owing to a genetic (hereditary) difference between his mother and his father in respect to any trait, the members of the pair of chromosomes carrying the determiner for the trait differ, then the new germ cells that are formed in that individual's body are of two kinds, one carrying only the mother's kind of determiner and the other only the father's kind. This is called segregation of the parental determiners in the germ cells of the progeny. Since the two parents usually differ not in one trait merely but in many independently inheritable ones, the individual germ cells are not of two kinds merely but of four or forty (or on the principle of combinations and permutations) 400 or more. This is the reason why the children of one pair of parents are so dissimilar, especially if the grandparents were unlike. This is the essence of Mendelism.

Varieties in the Human Species

The biologist has to recognize that the species *Homo sapiens* is a very variable one, just as its nearest living relatives among the different anthropoid apes are exceedingly variable. This variability is undoubtedly due to the fact that numerous mutations, or hereditary variations, have appeared in the human stock in the last 25,000 years and many of these persist in various parts of the world to the present time. These mutant traits are of various kinds, physical, mental, temperamental, instinctive. Of physical traits we have examples in dark brown skin, white skin and yellow skin; straight hair, curly hair, woolly hair; dark brown eyes, blue eyes; tall stature, short or stocky stature; hairiness and glabrousness; long head and short head; broad nose and narrow nose; thick lips and thin lips;

protruding brow, protruding forehead; protruding jaws and retreating jaws; teeth resistant to decay and teeth susceptible to decay; resistance to general infection and nonresistance; resistance to special diseases, like tuberculosis, cancer and to other diseases and nonresistance. These can be shown to be hereditary differences by the way in which they reappear in the progeny of particular matings and by our ability, in many cases, to predict what proportion of the offspring will show them. In their mental traits, likewise, different peoples are unlike. It has formerly been maintained that the obvious mental differences in races are due to differences of education and training merely, but experience with native tribes in Australia and in Africa has shown that the children of these peoples do not respond in the same way as the white children to the same sort of education. Also the army intelligence test, which was planned to measure intelligence that is independent of special training, showed that there is a marked difference in average mental capacity between the major races of mankind, and even between the peoples of different parts of Europe. We have abundant evidence to-day of an innate difference in capacity of learning, of forming judgments, of profiting by experience in different strains of humans. In fact, it seems probable that in the same country we have, living side by side, persons of advanced mentality, persons who have inherited the mentality of their ancestors of the early Stone Age, and persons of intermediate evolutionary stages.

The different races of man are unlike in temperament also. Some of them, like the Scotch, are prevailingly profound, thorough, somewhat somber; others, like the Mediterranean peoples, are prevailingly mercurial and lighthearted. The trait of reserve has been developed to a high degree among the North American Indians; that of fidelity to a superior race among the Bantu negroes; that of industry and dependability in the Chinese, and so on. Dif-ferences in instincts also there are between peoples. Thus the capacity for appreciating music has been shown to vary enormously and this variation to rest upon measurable dif-ferences in innate capacity for distinguishing such musical elements as pitch, time and intensity. Similarly, it is prob-able that there are innate differences in bravery, in aggres-siveness, in control of the sex emotions, and the like. It has, indeed, been denied by several observers that these differences are due to anything else than tradition and training; but there is, a priori, just as much reason for be-lieving that these differences in instinct and temperament are hereditary (racial) as there is for asserting a racial difference between a terrier and a collie dog in their reac-tion to rats or between a terrier and a chow in their temper-ament. Nobody doubts the hereditary nature of the differ-ence in instincts of the various races of dogs or of poultry and there is no more reason for doubting the hereditary nature of these differences in the case of man.

Man is commonly asserted to be a gregarious species and, therefore, a species which has had to devise *mores*, or social customs, for the organization of society. The *mores* of dif-ferent peoples differ; and they depend, to a considerable ex-tent, upon the instincts of the people. Peoples vary in the degree of development of the social instinct. Some peoples of southwestern Asia, for example, have the gregarious or social instincts relatively undeveloped. They are nomads, living together only in small groups of individuals that are more or less related. They are never builders of states or empires. On the other hand, the old Romans were organ-izers of states and cities, as are the Nordics of to-day. Again, one can hardly think of a religion which excluded music as arising among the South Italians, or a religion which rejected color and light from the church services as arising elsewhere than among the somber, philosophical

Scotch. However they have arisen, it is one of the principal aims of a people to secure adherence to its own *mores*. Such adherence is relatively less difficult with a homogeneous people, having the same instincts. Difficulties arise when alien peoples with different instincts come into the same communities and find themselves unable to meet the *mores* of that community. The eugenical ideals of a people demand that the population shall have hereditary capacity for meeting the *mores*. Good breeding, in the eugenical sense, is that which produces children capable of meeting the *mores*. Cacogenics, or bad breeding, results in the production of children incapable of the *mores*.

Since this world of ours contains many migratory peoples there is a constant coming in to any country of those whose instincts do not permit them readily to meet the *mores,* and society has attempted to meet this emergency by changing the instincts of the people through training. Some results have been obtained and, without waiting fully to inquire into their significance, society has come to regard training and education as the great prophylaxis for its defects and has poured a large part of its wealth into universal and uniform training. It may truly be said that society has, in general, been somewhat blind in what it is doing. It is perfectly true that by training and special culture the germs of innate capacity in different directions may be made to develop fully; perhaps they may be made to develop more fully and certainly more rapidly than without such special culture. But such special training and culture will not cause any capacity to develop in any individual who has not the hereditary germs of it in his make-up, any more truly than special training will change the temperament of a dachshund into that of a chow dog. Training, then, fails to secure uniformly good traits to a population of poor breed.

In any society a great variety of special capacities is required. These may involve mental attributes of a very different order, but each in its own way is important to society. Moreover, society requires not only persons with different mental, temperamental and instinctive reactions but also requires them in the right proportion. In the social insects not only have these proportions been worked out to a nicety but also the capacity for controlling them, and thus in any community of social insects we find soldiers and nurses and general workers in the right proportions. So human society is dependent upon its different kinds of workers and will suffer if it does not have the proper kinds in proper proportion during a great crisis. Thus during a great war it is necessary to have inventors, persons with the traits that are essential to successful aviation, to the handling of machines, to organization, administration, fighting. Any society which calls in vain upon these hereditary traits in its population would perish during a period of emergency which demands such traits. The great problem of eugenics is, then, to secure for any population the right proportion of the various capacities which that population requires, and for our particular civilization it is a question of eugenics to regulate these proportions under the *mores* of monogamy.

The Control of Matings

The first important factor in the problem is the control of matings. A proper mate selection implies, first, a knowledge of what society requires—what kind of traits and in what proportions; second, a knowledge of eugenics to get these hereditary traits into the population; and third, an ability to bring about the desired combinations.

If this problem, with appropriate specifications, were laid before a breeder of animals he could soon bring about the desired result in accordance with the specifications. The case is very different with humans and first of all because

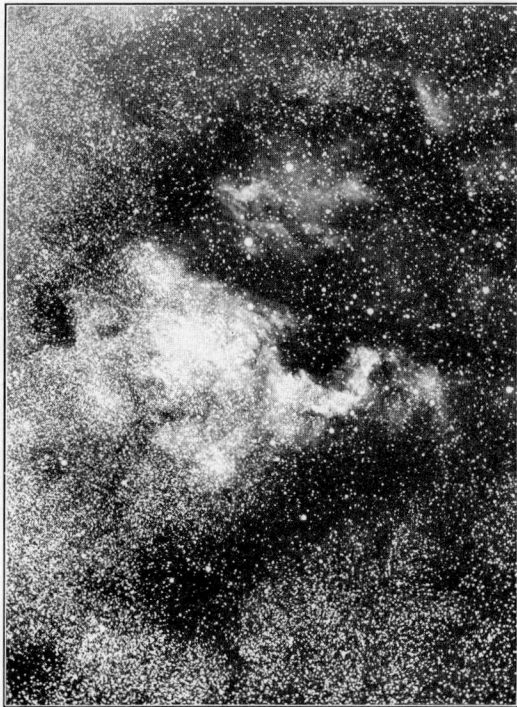

THE NORTH AMERICA NEBULA.

This is one of the vast gaseous nebulæ found in the Milky Way. Hydrogen and helium are present, but the principal constituent is the gas nebulium, characteristic of certain kinds of nebulæ, but not yet found on the earth.

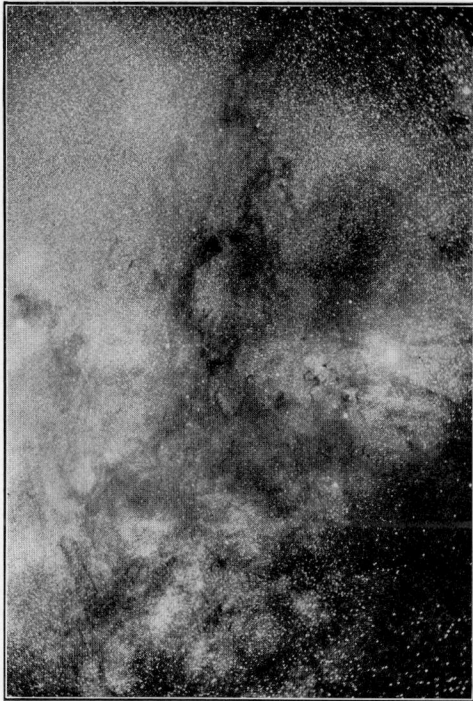

REGION NORTH OF THETA OPHIUCHI.

This extraordinary photograph was made with an exposure of 3½ hours by the late Professor E. E. Barnard with the Bruce telescope. It brings out some wonderful dark markings now believed to represent obscuring matter between us and the background of stars.

the desired result cannot be worked out by some superior being who has the matings of humans entirely under his control. We have, then, first to recognize the fact of human insistence on individuality and independent action—the difficulty in getting humans to subordinate their own behavior to the good of society. Indeed, it is a common human trait to show a certain resentment toward the wise, the superior, the elders. This is illustrated by the Greek voter who was opposed to Aristides because he was tired of hearing him called "the Just"; by the common laborer who constantly reiterates, with reference to his boss or some other superior, that he is "as good a man" as this superior. There is a constant tendency to tear down through individualism the social superstructure that has been slowly and painfully erected as a result of accumulated experience. This condition is painfully obvious to-day; it has been well set forth in Stoddard's book, "The Revolt against Civilization: the Menace of the Under Man." This difficulty has been accentuated by the constant encouragement of the inferiorly endowed by adherents of the false theory that differences between men are only differences in "opportunity." Such persons are blind to the fact that opportunities can never be made equal. There is a necessary inequality of opportunity because of a difference in inborn capacity for profiting by environmental conditions. If I am color blind I can never have an opportunity to learn to become a great painter. If I have little capacity for distinguishing musical pitch, the opportunity to become a musician can never be mine. If I am without the capacity for concentration, or imagination, I can never become a great inventor, or discoverer in science. There can be no opportunity imposed upon me in these directions from outside. That opportunity was lost at the moment I was conceived. Inborn, uneffaceable inequality of human beings is the one great fundamental fact of society, ignorance of which has caused the waste of untold millions of money in trying to make people alike and has resulted in a blindness toward the only method by which most social evils can be cured.

Another barrier to securing proper mate selection is again this individualism that insists upon following its own inclinations and refuses to be guided by the wisdom of elders. Various attempts to control mate selection by elders have been tried in different countries but have been far from universally successful. Thus in some countries, like France, it is the parents who attempt to control the selection of mates. In other countries it is largely the priests or patriarchs. We learn that in ancient Sparta, the state, even, took a prominent part in securing mate selection. In our own country, as in many others, the selection of mates is left to the persons primarily concerned. However, even in countries in which the young persons have the greatest freedom of selection the right of parental veto is commonly recognized.

But the case is by no means hopeless even when young persons are left to select mates without guidance from others; because taste and attraction are frequently, if not usually, directed by unknown elements which are in part eugenical. Thus the attraction of natural beauty has an important eugenic significance because beauty and full physical development, on the one hand, and hereditary health, on the other, are closely connected. One finds, indeed, that there are, all unconsciously, certain biological bases to selection. Thus persons of opposite temperaments tend, on the whole, to select each other and this is extremely important since the mating of persons of like temperament tends to exaggerate this temperament and leads to maniacal lack of control, on the one hand, or to melancholy, on the other. Again persons of similar stature and general build tend, on the whole, to select each other and this has an important eugenical outcome in preventing those physical, mental and nervous disharmonies which arise by a combination of unlike developmental tendencies in different organs of the body. It is very likely true that at the present moment the direction of matings by elders would be no more successful in the long run than that by the young persons concerned. Still, as a body of fact concerning the outcome of particular matings is gathered, it seems probable that teachers, physicians and parents, specially informed, may well be expected to exercise a greater influence upon mate selection.

MONOGAMY

Since the *mores* of our country favor monogamy, this is probably in accord with our instincts and capacities. It does not follow that it is the best system for all peoples, or for peoples with different instincts and capacities than our own. Polygamy is widespread in the higher gregarious species of animals and is a valuable method in the hands of the stock breeder. It is commonly practiced among the Turks, Chinese and other peoples. In these countries it probably results in persons of superior capacities becoming the parents of a larger proportion of the population than would be the case in monogamous marriage. The principal objection to polygamy, even in Turkey, is lack of domestic, especially paternal, influences. Again the suggestion has been made of a system of child-bearing without marriage by the method of artificial insemination which has been extensively used among domestic animals. From the standpoint of eugenics, this system is undesirable, since it does away with the very important social aspects of marriage, especially the providing of a place where the children may be trained, in the home under paternal restraint. From the same standpoint monogamy is far superior to promiscuity because it makes more nearly possible the construction of family pedigrees and a knowledge of the ancestry of each individual. Without a knowledge of matings it is impossible to make progress in eugenics.

Numerous as are the difficulties of securing appropriate mate selection, it seems probable that these are not insuperable. It is probable that, as we acquire knowledge concerning the results of particular matings and as this knowledge becomes widespread, it will influence matings, and as society develops under the monogamic *mores* and records are more completely kept it will be more and more possible for a pair of young persons to know with precision the probable nature of their offspring. The reason why mate selection is at present so haphazard and unscientific is largely because there has been no scientific basis for directing it; but just as with the development of the knowledge of hygiene its principles have become utilized in everyday practice, so it seems probable that as the principles of racial hygiene become established the knowledge of such principles will inevitably influence mate selection.

Human genetics is, indeed, the foundation stone upon which permanent progress of the race must rest, and although advance in our knowledge has been all too slow; yet some progress has been made in the past fifteen years. We know enough to predict the color of the eyes; of the skin and of the hair; the form of the hair; stature and build of the offspring. We can tell something about the probable special capacities in appreciation of form, color and music. We know something about the inheritance of capacities for resisting various diseases; something of the hereditary elements in the recurrence of particular undesirable physical or mental defects. We know something of the hereditary factors involved in the so-called functional psychoses, defects in emotional control and in social contacts. We know pretty well how the hereditary forms of epilepsy arise. We are learning the laws of inheritance of susceptibility to tumors. All this knowledge, however, is just at its beginning and requires to be refined and rendered more precise before it can offer a sure basis for mate selection.

DIFFERENTIAL FECUNDITY

The second great factor in eugenical progress is that of securing a differential fecundity. No matter how good, or bad, the mating, it has no eugenical significance unless there are children. If knowledge of the results of matings were known and widely followed, still there would be no improvement of the next generation if the well-mated had no, or few, children and those who mated without consideration of offspring were to continue to have large families. Eugenical progress would seem, therefore, to imply a fair number of children from the well-mated parents and no larger families, on the average, from those who are not well mated. The important social question is how to secure such differential fecundity as is implied in the high birth-rate of the best mated and low birth-rate from inferior stocks. One method which has been proposed is that of "voluntary" birth control, which is merely a spread of knowledge of the prevention of conception. This propaganda, which has been declared to have eugenical bearings, might have such if the knowledge were applied differentially; that is, if only the fecundity in non-eugenical matings were limited by the application of the methods of voluntary birth control. Unfortunately, however, it is precisely the less thrifty and the less foreseeing—the proletariat, or proliferators—who have now, as they have always had, the large families and who do not care for the principles of "birth control" any more than the rabbits do, and for the same reason—because they prefer to follow their instincts. Incidentally, too, they are well aware that society will care for any surplus of children beyond those that they can care for themselves. On the other hand, it is just the careful and thrifty who have been trained to exaggerate the importance of conditions of life upon the development of the children, who eagerly desire knowledge concerning "birth control" in order to improve conditions first of all for themselves, and, secondly, for such child, or pair of children, as they may have.

Organized society has occasionally attempted to control fecundity; especially to restrict that of inferior stock. Thus, segregation of defectives during the child-bearing period, which is now widely practiced in the United States by state authority, acts to reduce the fecundity of cacogenic strains. The principle of sterilization has been adopted by over a dozen states in the Union, but officials have been slow in the application of the principle because it seems to be in advance of public sentiment. No doubt, under proper control and used conservatively, the principle of sterilization by the state might become an important factor in differential fecundity.

All social attempts at controlling the fecundity of different races have been negative and little has been done toward securing larger families of the better endowed. In different countries, at different times, special privileges, or exemptions from taxation, have been given to those with large families. However, such attempts to increase fecundity have not been differential, but have applied equally to all classes and, it would appear, have probably been most utilized by those whose progeny is of the least value to the state.

DIFFERENTIAL MORTALITY

There is still a third factor in practical eugenics—the factor of differential mortality. It is not sufficient that matings should be eugenical and that the better matings should have a larger fecundity than the inferior matings.

It is necessary that these children of the better endowed should be reared to maturity. Even if the families of the better endowed were smaller than those of the poorly endowed, if a much larger proportion of them were brought to maturity, then the well-endowed, but relatively unprolific, strains might become in the next generation a dominant factor in the population. All governmental attempts to control mortality have hitherto been negative merely—directed to the prevention of deaths. They have been entirely undifferential. Equal attempts, if not greater, have been made to preserve mental defectives and persons without resistance to tuberculosis or other chronic ailments, than to preserve in health and productiveness those who are better endowed with good mental and physical health and are more resistant to disease. Anybody who visits an institution for the feeble-minded and sees the hundreds of "children" who are bedridden for life, who are paralyzed, incapable of speech, often with one or more senses defective or absent, and who are being fed and cared for as though they were infants through many years by effective, well-endowed attendants, must be impressed by society's failure to make use of one of the most valuable means that nature has provided for purifying the race. Even physicians who have it in their power, by withholding the exercise of their art, to permit the exitus of grossly deformed babies at birth refrain from using their right and duty of non-interference with natural death. Society is largely controlled by a perverted instinct—an instinct which is appreciative of the desire to live and blind to the value of death.

That this instinct is perverted we see if we try to imagine the consequences of a world in which there were no death, in which the people of earlier generations piled up, in ever increasing numbers, with all of their somatic damages, the results of accident and disease accumulated through life. It is rather remarkable that normal society has made so little use of the death function to free itself of its acquired burdens of physical and mental defectives, though if the less well-endowed gain ascendency, they use it freely to stifle competition. At the present time the state uses this method only in the case of capital punishment for murders and other heinous crimes, and there are persons who decry the exercise by the state of the method of death, even in these cases. On the other hand, everywhere we hear the propaganda for a low birth-rate and a low death-rate by persons who are blind to the effectiveness of nature's method, which is just the reverse, namely a high birth-rate, combined with a high death-rate. Above all, by this latter method, in all other species than man has permanent advance been secured. Man is trying the opposite experiment, and it is certainly failing. Progress cannot be made by the low birth- and death-rate method. The more this principle is exercised the more and more it will have to be exercised in the future in the attempt to stop, but all in vain, the rapid increase of the weak. Disaster will, doubtless, eventually overcome, in the future as it has in the past, any society that relies on this method, and its place will be taken by another more primitive society which at the moment is not controlled by such perverted ideals of social progress. .

Nature will, however, make one final attempt to save the strain, for during one period of the individual's development she may exercise all unhindered her beneficent method of purifying the race. This is in the period of fœtal development. It is at this period that the "lethal factors," of which geneticists have learned much in the past few years, play their part, inducing a partial or complete sterility, natural miscarriages and still-births. Thus, the weakest are killed before they are born. In one way or another nature will probably win over man's interference; the species will be preserved; if necessary, at the sacrifice of civilization.

SELECTIVE IMMIGRATION

There is one other factor that plays an important part, especially in our country, in determining the quality of the population. Even if all matings were eugenical and produced children capable of living to maturity, still the quality of the population might not improve if a large proportion of the population migrated into the country from other lands where eugenical principles were not at work. There have been years in the United States where the number of immigrants was not far inferior to the number of additions to the population by birth. Hence, the great importance to our country of exercising some control over immigration. The North American continent offered three centuries ago a clean slate upon which might have been written a glorious history of civilization. The immigration to certain sections of this continent from Europe was of the very best and the result was a rapid and solid development of the colonies in the first century of colonization. However, human greed and indolence led to the introduction of scores of thousands of a people among the lowest on the globe, namely of the negroes of Central Africa, and the blood of these peoples has mingled with the Anglo-Saxon and to-day is infused in about one-tenth of the whole population. Later, the greed of men and the desire for rapid wealth led to the introduction of thousands of the poorest peoples of Europe, until now our population has become in certain of our cities as mixed and of as low a type as that in some of the cities of the Balkans. Nevertheless, there are those to-day who still call for more cheap immigrants in order that they may enlarge their factories and pay larger dividends. America has nearly been ruined through the blindness of her people to the consequences of introducing inferior alien stocks. We have apparently been thoughtless of all consequences excepting the opportunity to gain greater wealth or comfort for ourselves.

Yet the whole matter of selective immigration is in the hands of the state, and the state has always exercised some control. At present we have laws against the immigration of feeble-minded, insane, criminalistic and the diseased. These are good so far as they go, but they are sometimes difficult to execute and they disregard entirely what is of great importance, namely, the family traits carried by the immigrants. Two recently-married immigrants may pass the physical and mental tests and yet carry in their germ cells such defects that some of their children are pretty certain to be defective. The only adequate basis of selective immigration is family selection, and we must have some knowledge of the family qualities of the immigrants. If such knowledge cannot be secured, then it were better to stop all immigration. Naturally the manufacturing industry will suffer at first, but later a better quality of persons capable of working with machinery will be attracted to manufacturing by the allurements of higher wages and the result may well be a larger product at no greater cost. It is not necessary that our factory hands should be feeble in mind and in inhibitions.

In this chapter I have tried to show ways in which through the utilization of precise knowledge the quality of our population may be improved, our social ideals realized, our permanent social progress assured. We may call the method that of eugenics, or we may use any other name, but in some way or other a nation which is laying the foundations of permanency and of social health must build upon the bed-rock of knowledge of the rôle played by the determiners of human traits which are carried by the germ cells in their chromosomes. The chromosomes!

The ignorant and the silver-tongued will scoff at these latest "figments of the imagination" of the highbrows; sentimentalists will cry out "for shame" at the material-ism of science; euthenists will call for more millions to mend conditions that are ever becoming worse. Yet to preserve or improve the nation there must be an apprecia-tion of the all-importance of the hereditary elements which nature provides free for those who know how to use them. Good heredity alone will give permanency and health and effectiveness to a people.

CHAPTER XIII

WHAT SCIENCE HAS DONE FOR AGRICULTURE

By E. DAVENPORT [1]

NEXT to religion and his own destiny, agriculture is man's greatest coöperative enterprise with nature. As a business it provides the food of all the people, the raw material for their clothing and other textile needs, and, until recently, it afforded the only means of land transportation aside from the labor of man himself.

It employs the time and talents of a full third of all the population, and by its refinements and perfections through the aid of science it has released the other two-thirds from the daily hunt for food that absorbs all the time of primitive man and holds him firmly bound in savagery and always on the verge of want. So have the blessings of art, education, refinement, comfort and security multiplied with the development and perfection of modern agriculture through the applications of science.

WOMEN THE FIRST FARMERS

Agriculture is as old as the race, for it began as a supplement to the hunt, when the women learned to scour the forest for nuts and fruits, and even the lowlands for the seeds of the larger grasses with which to assuage the pangs of hunger when the hunt should prove unsuccessful. To clear away competing vegetation from favorite plants, even

---

[1] Dean Emeritus of the College of Agriculture, University of Illinois.

to sow choice seeds in the new-made drift of the river bottoms, was as natural as it was to gather winter stores against a day of need.

In this way women were the first farmers, gradually bringing into cultivation a long list of grasses, shrubs and trees, each yielding something good for food just as the men domesticated the horse and the dog to help them in the hunt and later accumulated flocks and herds to assure a supply of meats and hides. So was the art of agriculture established out of nature's materials and so did it make possible a denser population than before, and an assured food supply with leisure to develop the arts and graces of civilization.

ADVENT OF SCIENCE

Agriculture as an art served mankind well for many thousands of years, but since the active advent of science about the middle of the last century, farming has undergone a development never dreamed of by our forefathers and altogether impossible under the superficial methods of our prescientific days.

Now, there is nothing uncanny or mysterious about science, nor is there anything requiring explanation or apology. Objectively, science seeks to find all the facts about a matter and to separate those which operate as causes from those which appear as effects. Subjectively, science is a state of mind trained to reason from the known to the unknown and interested only in truth, obeying literally the injunction of the scriptures—"Know the truth for the truth shall make you free." Science, therefore, is nothing but accurate knowledge combined with trained methods of thinking. With this knowledge and this training, sound conclusions and safe procedure may be worked out from which results can be predicted with a high degree of probability, even with certainty, provided all of the important facts have been discovered and correctly assessed, because the laws of nature are immutable and, therefore, entirely trustworthy. Without the scientific method of observation and reasoning all sorts of absurdities are possible, indeed inevitable.

For example, our grandmothers learned that a heated horseshoe dropped into the cream would often help to bring the butter, but, not knowing all the facts, they ascribed results to the driving of witches out of the churn, whereas the butter came because of the increased temperature, and any source of heat would have been as effective as was the traditional horseshoe.

Just as churning, which was one of the difficult arts of former times, has been reduced to scientific accuracy by the use of bacterial starters and correct temperatures, so has science everywhere freed agriculture from a vast mass of hampering traditions which were the accumulations of generations of unscientific attempts at coöperation with the laws of nature.

It was not in the ancient world but in our own America that opposition to the iron plow developed upon the ground that the metal of the share would poison the soil, an assumption without the slightest evidence, and in face of the fact that iron is everywhere present in the land and is absolutely essential to both plant and animal life.

It was not merely the men of the Old Stone Age who consulted "signs" before undertaking important operations, but it was our own immediate forebears personally known to many of us. But thanks to science most of this popular befogment has been already driven away like mist before the rising sun.

In the serious business of producing crops to feed a dense and hungry population, it is difficult enough to secure good seed of suitable varieties, plant it in well-prepared and fertile ground of the proper degree of moisture and protect the plantings from weed, insect and fungous enemies until the harvest, without feeling obliged to perform all these operations with regard to the phases of the moon, the signs of the zodiac, or the dreams of the farmer and his family.

THE SCIENCES THAT SERVE AGRICULTURE

Though all truth is helpful, yet five great fundamental sciences lie at the basis of successful farming and stock raising, namely: chemistry, biology, physics, physiology and economics.

They came into its service in about the order named, chemistry first, beginning about an hundred years ago and exceedingly active for about half the intervening period, with biology dating from the middle of the last century and prolific in results for the last generation, followed by physics, physiology and economics, each shedding more light every day upon the business of producing crops and live stock.

Broadly speaking, the great enterprise of agriculture is the growth of crops for the food of man and his domesticated animals. Even though we consume a great deal of flesh, milk and butter, yet the animals that produce them live upon crops and, therefore, in the last analysis, in civilized life as out in nature, all animal life, man included, lives upon plants, whose business it is by the aid of sunlight to elaborate highly organized materials out of the simple elements and compounds of soil and atmosphere. Scientifically, therefore, the basic question in agriculture is first to know what man requires successfully and fully to nourish himself, then how to produce that food continuously and economically out of the soil without impoverishing the land and thereby cutting off the food supply of succeeding generations.

We know now that the function of food, whether of man

or his animals, is two-fold, namely: to furnish materials for the building and repair of the body, which is chemical; and to provide the energy or power for doing its work, which is a matter of physics.

Instead of the body being built like a machine in a factory it builds and repairs itself, as it goes along, out of the materials of its daily food. Not only that, but the energy or power necessary to this building or repair and to enable the body to perform its daily work external or internal comes not from a boiler or an electric motor, but is liberated from the breaking down, into simpler combination, of the same food whose atoms and molecules provide the materials for body building and repair.

In other words, we take food not merely, or even mainly, for the materials it contains, but we literally stoke our bodies for energy precisely as we stoke our boilers with coal in order to make steam or electricity, and too many people are weak and listless only because they neglect or refuse to provide themselves with sufficient energy to meet the demands of civilized life, either by withholding food or by neglecting to take in sufficient oxygen by deep breathing to liberate all its power, and make its materials fully available.

How Crops Work

Science has shown us that the growth of crops is a synthetic process by which all plants build up out of the atmosphere and from the soil great quantities of the so-called hydrocarbons, compounds of carbon and of hydrogen and oxygen in the proportion to form water. Besides this "starch," "sugar," or "cellulose," which make the bulk of the plant, each species manufactures out of the materials of the soil small quantities of one or more peculiar and exceedingly complex compounds nearly always containing nitrogen and generally some other of the ordinary chemical elements, especially lime, phosphorus, potassium, iron and sulphur. These constitute the special flavors of fruits and vegetables, as well as forming many of the fats and other products of peculiar properties, even medicinal.

For example, there have been discovered only recently what are called vitamines, two or three of them, which seem to be needed in small amounts, especially in young and growing bodies. Until recently white and yellow corn were supposed to be of identical feeding value, which is correct as regards ordinary constituents. But while rats and pigs may be readily raised on yellow corn they languish and fail to develop when fed exclusively upon the white varieties, the difference being a minute amount of one of the vitamines contained in the oil of the yellow corn. Differences such as these are the cause of feeding babies these days upon such unaccustomed materials as orange juice, tomatoes and spinach.

These rare and unusual compounds formed in both plant and animal tissues are due in every instance to that subtle alchemy which is peculiar to the species. Broadly speaking, therefore, all the farmer has to do is to provide the general conditions of growth and protect the crop from its enemies until maturity. Nature will do the rest.

So far as the energy needs of the body are concerned, we know now that while the carbon compounds are *par excellence* the energy foods, yet all complex compounds will furnish energy to the body as they are broken down into simpler compounds following digestion. This liberated energy is then available first, to do such internal work as pumping the blood over the body, manufacturing peculiar animal products such as milk, body tissue, or glandular secretions; second, providing the power for walking, running, or performing any kind of labor that may be put upon the animal. Any surplus energy—and there is usually a surplus—will escape as heat of low intensity, as will also all forms of arrested motion such as muscular contraction. This accounts for the phenomenon that all living tissue is giving off heat, and is, in general, warmer than the surrounding medium, a fact that disproves the unscientific hypothesis that we eat in order to maintain the body temperature, the truth being that the body is warm because of its work, its temperature being the algebraic sum of all the escaping energy upon the one hand, and the accident of radiation upon the other. That is why different animals with differing food habits and different bodies of differing ages and shapes exhibit different temperatures.

The Fertility Question

Science informs us that in general about ninety-five per cent of our crop is made up of water and the so-called hydrocarbons coming out of the free supply of the atmosphere, and therefore beyond the farmer's concern or control except in irrigated countries where the value of the water right is frequently more than that of the land. This leaves but about five per cent of the crop as involving a draft upon the fertility of the land and often even less, for in the case of the apple it amounts to less than one per cent.

About half of this five per cent not water or hydrocarbons—more or less according to the species—consists of nitrogen compounds, the nitrogen of which comes ultimately from the free supply of the atmosphere, but must reach the plant through its roots and its leaves, and in a combined, not a free, form. The other half, more or less, consists of minerals that must come directly from the limited supply of the soil, and, therefore, this one or two per cent represents a real depletion of soil fertility and its power to produce crops.

Here, then, are two tremendous questions in crop production and soil fertility that have engaged the attention of farmers for generations, and of scientists since chemistry and biology discovered the two spots at which the art of agriculture had failed to solve its problems. These are the nitrogen question and the mineral question.

THE NITROGEN QUESTION

Farmers had long known that certain manures were good for crops and that certain plants were beneficial in the rotation, but they never understood the reason for it and the principle involved. Accordingly, when chemists began to analyze crops the farmers jumped to the conclusion that everything found in the plant should be supplied as fertilizer, and such absurd applications were made as sand to strengthen the straw of grains.

Chemists, on the other hand, assumed that only such fertilizing materials need be applied as came originally from the soil, namely the ash that remained after burning. They reasoned that as three-fourths of the atmosphere is nitrogen, the plant could look out for itself in that respect, particularly as the visible supply of that element is more than a million times as great as that of carbon, which makes the bulk of our crop and comes directly from the air. But while carbon unites readily, especially with oxygen, yet nitrogen, the "lazy element," is exceedingly difficult to entice into combination of any kind.

It was learned that the electric current of thunder showers was able to convert small amounts of atmospheric nitrogen into nitric acid, which, being brought down by rain, formed nitrates in the soil and as such could be taken up by the roots. But the amounts that could be so formed were insufficient to account for the needs of plants, besides crops were successfully grown in rainless regions. There must be, therefore, some natural means either of driving or teasing this stubborn element into combination! But for many years it eluded the best efforts of chemist and biologist alike until the "nitrogen mystery" became the scientific puzzle of the later years of the nineteenth century, during which time England exhausted the guano beds of the South Sea Islands that had been thousands of years in forming, while scientists and farmers alike talked of "nitrogen starvation."

Finally, as late as 1886, the mystery was solved. It had long been noticed that certain pod-bearing plants, which the botanists called legumes, bore on their roots certain warts, or little nodules, which some regarded as a disease to be bred off. But the microscope revealed these tubercules as colonies of bacteria or microscopic plants which have the remarkable power of utilizing the free nitrogen of the air and combining it for the use of other plants, even their own hosts, clover, alfalfa, cow pea and other pod-bearing vegetation, whether succulent herbage like the clovers, or trees like the honey locust and catalpa. The nitrogen mystery was solved, the value of clovers and other legumes in the rotation explained, and it has since been noticed that the first vegetation to spring up on coal dumps, old lava beds, mountain sides or the refuse from deep excavations is some species of lowly legume with its roots covered by nodules, each a colony of laborers, who "work for nothing and board themselves."

The limitations of space forbid pursuing this fascinating phase of agriculture further except to say that each species of legume supports its own peculiar variety of bacterium, and while the legume, like other plants, will grow well in rich soil without the little parasites, yet if the land is poor it must be supplied with nitrogen by its own peculiar breed of helper. So true is this that we now prepare separate inoculation for clover, alfalfa, cow pea, and all the agricultural legumes. The nitrogen mystery is solved, and the nitrogen question has become merely the one of introducing

sufficient legumes into the rotation to keep up the supply
of combined nitrogen for the use of the other and naturally
exhausting agricultural crops.

THE MINERAL QUESTION

The ash of plants contains in small amounts two classes
of minerals, namely: such as may happen to have been
taken up by the roots because dissolved in the soil waters,
and therefore accidental; together with certain others which
are as essential to plant growth as are nails to the construction of a building. Of these necessary minerals two
at least are likely to be deficient in agricultural land and
to become exhausted by continuous cropping, however
abundant the others may be. These are potassium,
necessary to the successful growth of leaf and stems, and
phosphorus, of special need in the ripening of grain.

Every crop removes definite amounts of these necessary
elements. An hundred bushels of corn, for example, takes
out of the soil no less than seventeen pounds of phosphorus
and nineteen of potassium, while a hundred bushels of
wheat will remove twelve and thirteen pounds respectively,
or sixteen and fifty-eight if the straw is sold as well as the
grain.

The farmer gains much by feeding his roughage and
returning the refuse back to the land, but in spite of the
best practices there is a small, but continuous subtraction
of the fertility of the soil, and its power to yield crops
unless these two minerals, so indispensable to plant life, are
systematically brought from the mines and applied to the
soil by main strength.

The physiological fact is that, no matter how well supplied a soil may be with most of the conditions for growth,
the crop will be limited by the supply of the least of the
necessary ingredients just as the construction of a building may be limited by the number of nails, bricks, or other

needful material, with this difference, that the crop will
not hold up the job until sufficient material is at hand to
insure a full completion of the plans, but it will go ahead
with the enterprise at once and continue as best it can until
the limiting element is exhausted.

This limiting element in the land is generally nitrogen
or phosphorus. Of the two, nitrogen is far the more ex-
pensive yet the easier to obtain by the methods outlined,
for more than twelve pounds of it rest on every square inch
of every field, and if only the farmer will set his little
bacterial neighbor at work the nitrogen need not be brought
from the South Sea Islands.

THE UNSOLVED PROBLEM IN AGRICULTURE

The problem of a permanent agriculture is therefore the
problem of keeping the mineral supply up to the demands
of maximum crops. This is scientifically possible by methods of direct application now well known, and in successful
use. It is economically possible with the good farmer, but
it is psychologically impossible with the shiftless or im-
provident man who neglects or refuses to look ahead, or
who is so much of a skeptic as not to believe what he cannot see with his own eyes, and who is deceived by the ups
and downs of seasonal variations in yield. Such a man
does not believe in failing fertility, even in bacteria, and
finds it convenient to say so as a cloak for his neglect.

The question therefore is whether the better class of
farmers can restore the land, generation by generation, as
fast as the poorer farmers will wear it out or whether
some sort of public control will have to be exercised to con-
serve fertility, upon the ground that if the food supply of
the land fails the nation will disappear. In any event
science has pointed out the natural principle involved; the
issue lies with the farmer and with society. The scientist
has made coöperation with nature possible.

CROP IMPROVEMENT

For thousands of years the crops of the farm consisted merely of wild vegetation brought under cultivation just as the domesticated animals were the products of nature brought in from forest and plain, tamed and trained to the uses of man.

But with a better knowledge of biological laws the possibilities of crossing suggested themselves. In nature the number of crosses that can take place, and therefore of new varieties likely to occur and survive is exceedingly limited, but in cultivation the possibilities of new strains through new combinations is almost unlimited, and we have only to note the new and vastly improved varieties of garden fruits and vegetables as compared to anything the world knew a thousand or even a hundred years ago in order to realize that science has been hard at work in producing new and valuable varieties for the use and enjoyment of man.

Strawberries, blackberries, raspberries, and tomatoes are new, as is head lettuce, grapefruit, and long staple upland cotton. Almost a multitude of new and strange creations among the flowers have set the world ablaze with glory. Man has made none of these from wholly new materials, but he has so remade what nature spontaneously produced that the new can hardly be recognized as connected with the old.

The most valuable fact in all this exploration into the realm and mysteries of heredity is this: science has shown us conclusively that heredity works by definite units that are as concrete, and as certainly capable of identification as are the bricks in a building. Not only that but in all bisexual reproduction whether of plant or animal, these units from the two parents combine in exact mathematical ratio and according to the binomial theorem. In other words, if the breeder is sufficiently well acquainted with his materials he can with confidence predict the results of peculiar combination of blood lines and can also with confidence foretell the proportions of possible varieties that will result from each particular combination.

PLANT BREEDING

With this advanced knowledge some remarkable work has been done in changing the natural constituents of plants as well as their external form.

For example, the time came when Germany desired to be independent of the tropics in the matter of sugar production, and set out to discover and breed up some sugar producing plant other than cane. The common garden beet looked promising, but its sugar content ran only three to four per cent. By continuously planting the best, however, raising seed only from the roots showing the highest sugar content, a beet was finally bred up that produced field averages of twelve and even fourteen per cent with individual plants running as high as sixteen. This is an achievement that was impossible before the days of skillful chemical analyses.

This change in composition is of the root and the question arose at the University of Illinois some years ago whether the natural constitution of a seed could be similarly altered. Dr. Cyril G. Hopkins undertook the task with the result that after nine years of selective breeding for increase of nitrogen, the protein content of the seed was raised from an average of 10.92 per cent to 14.71 per cent, or higher than that of wheat. Selection for oil content was carried both ways from seed that averaged 4.70 per cent. The spread between the high and the low strains was noticeable at once, and the ninth crop produced an average low oil of 2.58 per cent, and an average high oil of 7.29 per cent.

That this is due to progressive improvement, and not simply to assortative selection is shown by the fact that with each successive crop individual ears appeared beyond anything that was to be found in the original stock. It is further shown by the fact that progressive changes are still going on, for in the twenty-sixth crop the protein content has risen to 17.33 per cent, while the spread of oil is from 1.76 to 9.86 per cent.

The limit will evidently not be reached as long as the changes do not impair the vitality of the seed, and these experiments show not only the flexibility of species, but the wide margin of safety with which nature provides her children.

PLANT DISEASES

Associated with every valuable crop are certain weeds that so nearly imitate its habits as to become a more or less serious handicap in its production. Weeds, however, can be seen by the naked eye and, however troublesome, will disappear with thorough cultivation.

But there are crop enemies that work in ways not met by cultivation. There are rusts, smuts, and blights; rots, molds and scabs; locusts, worms and lice; bugs, beetles, and butterflies; all under the general head of fungous and insect enemies. These all prey upon the crop, either by eating its foliage like worm and beetle, by sucking its juices like the bugs, or by actually penetrating its substance, and living off the tissues like the rusts and smuts.

These enemies all multiply prodigiously, indeed nutrition and multiplication are their specialties, and unless successfully combated they will either ruin the crop or so reduce the yield as to destroy all profit to the farmer and menace the food supply of the people.

Science first attacked the worms and beetles that chew the leaf and destroyed them by poisons so thoroughly sprayed as to reach every portion of every leaf. It next learned that fine powders would kill many injurious insects by clogging the breathing pores, and finally it discovered that certain repellents were effective, while irritants like tobacco, and kerosene solutions would destroy many tender insects and most species of fungi by contact. By one or all these devices our fruits and vegetables are now fairly well protected against a host of parasites far more terrible than an army with banners, for almost any one of these enemy species is entirely capable, if unhampered, of taking complete possession of the earth within a generation.

ANIMAL DISEASES

The domesticated animals, like the cultivated crops, are all subject to the attacks of parasitic enemies. Such external pests as lice and ticks are easily removed by poisoned dips, but not infrequently these blood-sucking insects cause infections of most serious internal disorders, such as Texas fever.

For quite aside from the pests of insects there is a long list of animal diseases due to the invasion of low forms of bacterial and protozoan life that attack our domestic animals as they do us, and many of them, like anthrax and diphtheria and tuberculosis, are common to both.

If distinctions can be made in so disreputable a company in which anthrax, contagious abortion, foot and mouth disease, and Texas fever are shining examples, it would be said that the two most common and most troublesome of all the animal diseases are hog cholera and tuberculosis, especially the latter because it is so generally distributed, and so closely identified with the same disease in man.

There is no longer doubt that tuberculosis passes back and forth reciprocally between man and his cows, just as we know that cats harbor children's throat diseases, especially diphtheria.

Laymen have fought the scientist in his conclusions as to the practical identity of human and animal tuberculosis, just as has sentiment for pets harbored the cat in most intimate family relationship with children when it is perfectly well known that she is a nocturnal prowling wild animal, living largely off birds which are not only the farmer's best friends, but a joy to all lovers of nature. Science has come out ahead on tuberculosis, but we still harbor the cat, which shows that sentiment is even stronger than knowledge and judgment combined.

The "Great White Plague" is not a visitation of God, but a parasitic invasion that can be not only controlled but exterminated. It is due to a minute and rather delicate organism that thrives in the tissues of man only at times of fever, but finds congenial habitat in cows whose temperature is over three degrees higher than ours and between them, and us, the disease readily crosses over, milk being the vehicle from cow to man, and promiscuous expectoration about barns offering the return infection from man to cow. In pasteurization, science has practically closed the door against infection by milk.

The same disease attacks birds, but their temperature has produced a special strain requiring a high temperature, and therefore not easily communicated to man; besides, the flesh is cooked before eating, a process of most effective sterilization.

IMMUNITY

Strangely enough, horses are immune to tuberculosis, just as certain individuals are immune to other organisms. Not only that, but many infectious diseases such as small-pox are endured but once because the body seems able while suffering from the invasion to manufacture a high degree of immunity which finally destroys the parasite, and is more or less permanent in its effects.

These facts have led the scientist to experiment on artificial immunity through serums and latterly upon natural immunity by breeding from immune stock. It is too early yet to speak with more than hope as to hereditary immunity, but a long list of serums has been prepared, each almost a specific against its special disease. In no field of endeavor has science been more successful or more beneficial than in its combating of diseases, especially those due to invasion from without.

MACHINERY

Finally, science has done much for agriculture along engineering lines by the development of such machinery as the world has never known before, and this is the principal reason why more than half our people are released for work other than food production. The first harvesters were burned by laborers because they feared the loss of employment, but these laborers are now making automobiles and better furniture than ever before, as well as a thousand conveniences for common people that even kings in the olden time never dreamed of.

By machinery the American farmer works many times more acreage than any other farmer of the world works or ever did work, and does it better by far than hand labor is capable of doing.

We now plow, and plant, and cultivate, and spray, and harvest, yes, pack and sack by machinery; indeed many a crop is produced almost literally without being touched by the human hand. And the end is not yet, for new machines are coming out every year whereby the strength and handicraft of a man are multiplied many times. All this has not only lightened the labor of the farmer and cheapened food to the consumer, but it has vastly enriched our arts and our civilization by releasing so large a portion of the people for labors other than food production.

The advantages of mechanical labor are not confined to the farm, but are extended to the home. The separating of cream from the milk used to be a long and difficult undertaking requiring crocks and pans by the score. The labor is now better done in a few minutes by mechanical separators. Churning was, until recently, a prolonged and difficult labor with great uncertainty as to results. Now, with bacterial starters and power machinery, the process is reduced to a matter of minutes with absolute certainty as to the product.

With the perfection of the internal combustion engine, we now have perfectly operating private plants delivering water under pressure, making electric lighting for house and barn, driving not only separator and churn, but washing machine and mangle.

Last of all is mechanical refrigeration, just emerging from the experimental stage, but proving entirely feasible. All this machinery is perfectly automatic, needing little or no attention except for lubrication and providing the fuel supply.

WATER CONTROL

As about seventy-five per cent of all crops consist of water, and as many times as much water must be evaporated from the leaves as is retained by the plant, it is easy to see that water is the limiting element in crop production.

It requires a total of about twelve inches in depth over all the surface to grow and mature the average crop. This is independent of what is evaporated from the surface of the soil without going through the leaves, and must all be provided during the growing season.

The annual rainfall ranges from nothing to a little over twenty-one feet at Greytown, Nicaragua, which seems to hold the world's record. The interior of all continents is arid, not to say desert, while agricultural regions generally enjoy about thirty to thirty-six inches of rainfall, though dry farming is carried on with only about half that much where the distribution is favorable and the surface evaporation not excessive.

In all rainy regions the run-off constitutes serious agricultural and engineering problems, and even in arid regions the infrequent rains often descend in torrents that cause washouts and landslides that vastly change the face of nature with every generation.

The problem of excess water, therefore, calls not only for protection against floods on the lower levels, but also against erosion and the carrying away of large quantities of the better soils from the uplands off to sea, for it is only the finer portions of the earth's surface that constitute soil within the agricultural meaning of the term, the great bulk of the earth's materials being entirely inert.

When we realize that the mouth of the Mississippi was at one time at Cape Girardeau, Missouri, and that the great valley below has been filled for a thousand miles by silt brought down and dumped into the Gulf of Mexico; when we look at the delta of the Nile, the Hoang Ho, and every other great river of the earth; when we recall that the silty, water-made island of Marajo at the mouth of the Amazon is as large as Ireland, or when we reflect that no less than twenty per cent of our prairie states are subject to erosion, then it is that we realize the magnitude and the seriousness of our agricultural-engineering problem of water control.

In the United States we have more than a hundred rivers, big and little, and all are highways for excess water, sometimes needed for irrigation, often a source of danger, but always a source of power.

More and more, therefore, are engineers and agriculturists becoming convinced of the need and utility of what might be called a three-way control—first, impounding all excess water as protection against floods; second, using it for irrigation whenever and wherever needed; third, utilizing its power as it is gradually liberated, whether used for irrigation or not, the hydro-electric value generally being enough to repay the cost of control. The really serious phase of the problem is the area needed for impounding, particularly in the more level and valuable agricultural regions; a problem that does not exist to any great extent in the foothills.

THE FUTURE

In the fifty years or less of its activity science has only begun to show the character and the extent of the service of which it is capable, and almost before the last achievement can be put into print some new and unexpected discovery has become a working reality. It is too early even to discuss the possibilities of radio activity, but that it will exert its influence in unexpected ways in agriculture is not at all improbable. The time has come when one dare not attempt to predict what will happen next and certainly no careful student will assert what will not be done by science.

Dr. Slosson in a recent number of the *Scientific Monthly* quotes the eminent French scientist, Haiey, as saying at the beginning of the nineteenth century: "Electricity, enriched by the labor of so many distinguished physicists, seems to have reached a time when a science has no more important steps before it, and only leaves to those who cultivate it, the hope of confirming the discoveries of their predecessors, and of casting a brighter light on the truths revealed."

In the late seventies one of the distinguished chemists of the day said to the writer that while electricity is an exceedingly interesting and a very powerful form of energy, he doubted if it would ever be of any practical use beyond telegraphy. This man lived to use the telephone and there are those who believe that the electric current, or some similar subtle energy will yet be harnessed, and put at work at combining nitrogen for agricultural purposes. The science of it is already worked out and the only question is one of economics. Should this be the turn of the nitrogen question it would vastly alter the rôle of leguminous plants and work a revolution in crop rotation.

Whoever soberly considers what science has achieved for agriculture in the short space of half a century, can but render thanks to Almighty God for His revelation of the laws of nature, and he will face the future with confidence unlimited and with gratitude unbounded.

CHAPTER XIV

SANITATION

By C.-E. A. WINSLOW [1]

HISTORICAL DEVELOPMENT OF THE PUBLIC HEALTH CAMPAIGN

THE public health campaign, in the sense of an organized community movement for the prolongation of human life, and the promotion of physical vigor and efficiency, is a relatively recent phenomenon in the history of the human race. Personal hygiene achieved a brief flowering in Greece, municipal sanitation in Rome. The principles of isolation and quarantine proved effective in controlling leprosy, and were applied with less success in the attempt to control plague and typhus fever in medieval times. It was, however, the work of Edwin Chadwick and John Simon in England, beginning in 1838, which ushered in the Great Sanitary Awakening. This movement was at first based on incomplete theories as to the causes of disease, but its instinctvie dread of filth was fundamentally sound. It initiated a movement for the cleaning up of the material environment, for sewerage and sewage disposal, and for the protection of water supplies, which relegated some of the worst of the plagues of the Middle Ages to a position of purely historical importance.

Meanwhile, in France, Louis Pasteur, between 1860 and 1880, was laying the foundation for a really scientific knowledge of the causation of disease. He demonstrated

---

[1] Professor of Public Health, Yale School of Medicine.

that the communicable maladies were due to microbic parasites, and he taught us the science of bacteriology by which these microbes can be detected and destroyed, and the science of serology by which the vital resistance of the human body can be built up to resist their attacks.

The first phase in the development of the public health campaign was environmental sanitation—largely an engineering problem—and the second was concerned with the control of community infections by bacteriological methods. As these procedures began to yield results a new set of problems was forced upon the attention of sanitarians, problems which required a new technique for their solution. The control of tuberculosis and infant mortality, for example, could only be attained in a relatively slight measure by environmental sanitation or by isolation and quarantine. The essential thing here was to alter the habits of the individual, to teach the principles of healthy living. The dominant note in public health work for the past twenty years has been the teaching of personal hygiene.

Since the particular applications of personal hygiene must vary widely in the individual instance according to the needs and the limitations of the individual, there has become manifest in recent years an increasing tendency to develop the applications of preventive medical service, which promises to be the outstanding characteristic of the coming decade. Infant welfare clinics, prenatal clinics, school clinics, tuberculosis clinics, venereal disease clinics, are being inaugurated for the medical examination of well persons or at least of persons not acutely ill—for the detection of incipient disease at a stage when it is controllable and for the hygienic direction which will maintain health. Thus, not only environmental sanitation and the control of epidemic disease, but education in personal hygiene, and the organization of medical service for the prevention of disease come in as essential parts of a rounded campaign which regards no disease but old age as beyond the possible range of preventive effort.

SCOPE OF THE PRESENT DISCUSSION

We are not concerned here with a discussion of the entire field of public health as outlined above, but with a review of a definite and limited section of it. The health of the individual depends upon three sets of factors—the inherited potencies which he derives from his ancestors through the germ-plasm, the environmental conditions which surround him, and the management of the living machine which develops from those potencies under the influence of that environment. The first set of hereditary factors is clearly delimited, and has been discussed in a preceding chapter. The last two overlap to some extent; and yet it is not difficult to draw a distinction between those broad external conditions which exert a generally similar influence upon all human beings and the special problems of hygienic conduct which must be adapted to the individual, frequently under the limitations of specific medical advice. For our purposes we shall consider that such fundamental requirements as the need for essential food elements and suitable atmospheric conditions form a part of the general subject of sanitation.

FUNDAMENTAL ENVIRONMENTAL NEEDS OF THE LIVING ORGANISM

If we attempt to analyze the conditions which are essential for the existence of a living being, from the minute bacterial cell to the complex organism called man, we find that they can be reduced to very simple terms. There are three fundamental requirements for the life process, water, food and the proper chemical and physical conditions of the surrounding medium; and there are two fundamental dangers which must be avoided, mechanical violence and the activity of other hostile forms of life, predatory or parasitic. So long as the first three essentials are provided and the last two dangers are avoided life may go on. It is to the discussion of these five conditions of living that we shall confine ourselves in the succeeding pages.

FOOD NEEDS OF THE BODY

In the case of the human being death from thirst is a rare and exceptional accident. It may be dismissed therefore with this mere mention, and we may pass to the second problem, the food needs of the body.

The human body is subject to the fundamental law of the conservation of energy in the same way as a steam engine or an automobile. Every step taken in walking, every flicker of an eyelid, every beat of the heart, as well as a thousand subtle processes constantly going forward within the body without visible signs of movement, with all those chemical changes which produce the bodily heat, must require the expenditure of energy; and this energy comes from the food. The ordinary unit of heat energy which we use for measuring this flow of force is called the calory, and is defined as the amount of energy which would raise the temperature of one kilogram of water by the amount of 1° centigrade. The minimum requirement for the maintenance of physical life reclining on a couch is about 1,600 calories per day, while the maximum may rise to 10,000 calories under such conditions of violent exertion as obtain in a six-day bicycle race. The first thing we must get from our food, therefore, obviously is this daily supply of energy.

It is by no means only energy, however, which we must obtain from our environment in the form of food, if we would live. By its very nature protoplasm, the living stuff of our bodies, is constantly breaking down, and this con-

tinuous waste must be as constantly made good by the supply of the substances out of which the body is built up.
We must have building stones as well as fuel, particularly
in childhood when repair exceeds waste and growth results, but with equal urgency if in less amount when the
growth period is over and intake and outgo merely balance.
To supply these necessary building stones is the second
function of the food.

We need then first of all to obtain from our daily diet
some 2,500-3,000 calories of energy, and this need can be
met by many combinations of various foodstuffs. One
hundred calories will be furnished for example by three
large lumps of sugar, one large egg, an ordinary serving
of oatmeal, one chop, two slices of bread, one orange or
two-thirds of a glass of milk. The cereals, pork products,
sugar and fats are the chief foods which meet the calory
requirements of the body. In addition, however, the diet
must be so balanced as to yield along with the calory allowance a supply of the special substances which the body
needs for the continuing repair of the living mechanism.
First of all it is essential to provide proteins or nitrogen-
containing foodstuffs in the proper amount and of the
proper kinds, for there are many kinds of proteins, and
some of them are by themselves lacking in some of the
necessary building stones. Meat, fish, eggs, milk and cheese
and certain vegetables like beans are rich in proteins. Next
we need certain mineral salts, calcium, iron, phosphorus,
and others which are found in milk, eggs, vegetables and
fruits. Finally we must obtain minute amounts of mysterious substances, called vitamines, whose nature and function are not clearly understood, but which are essential to
growth and to health, and which we obtain largely from
milk, butter, green vegetables and fruits.

The human machine may suffer from too much food as
well as too little, but this is purely a problem of personal

and individual hygiene. The provision of the materials for an adequate diet may without stretching our definition too far be considered a problem of sanitation. Malnutrition is a common condition, not merely in Russia or India in famine time, but among the school children of every city, and every countryside. It is due in some instances to an actual lack of the necessary foodstuffs at prices within the scope of the family purse. When such is the case there is a real problem for the statesman to stimulate food production, to improve food conservation, and food distribution, or to raise the economic status of the wage earner. More often perhaps malnutrition is due to lack of intelligence in selecting from those food sources which are at hand; and here the problem can be solved by education in dietary hygiene. The modern movement initiated by the Child Health Organization for the development of sound food habits among school children is a powerful force working toward this end. In the interest of a balanced and an economical diet the average American food budget should be greatly reduced in respect to the sums spent for meat, fish and eggs and increased in the amounts spent for milk, vegetables and fruits, since these foods supply the much needed salts and vitamines while milk is a cheap and ideal source of proteins as well.

The Atmosphere in Relation to Health

The third fundamental essential for the conduct of the life process is an enveloping medium of the right chemical and physical constitution; and in the case of the human being this involves the problems of ventilation or air conditioning. Chemically it is necessary that the atmosphere surrounding the body should contain the oxygen required for the constant oxidations which go on all the time in every living cell, and form the basic element in the life process; and that it should be free from toxic gases. Physically, the atmosphere must be of the right temperature and humidity, and must not be laden with injurious dusts.

Our conceptions of the practical criteria of ventilation have been radically modified by the investigations of the past twenty years, which have tended to emphasize the physical rather than the chemical problems of air-conditioning. Men may suffer from oxygen starvation in the deep parts of mines, and in old wells or cesspools, as they do in the rarefied upper air on mountain tops or in aeroplanes. In even the worst ventilated room above ground, and at ordinary levels, however, the amount of oxygen present changes but slightly, as a result of the ready diffusion which takes place through walls and ceiling. For practical purposes the danger of diminished oxygen content may be ignored as a ventilation problem.

Nor are poisonous gases by any means so general a danger as was once supposed. In certain industrial processes toxic fumes are produced which constitute a real menace, and which must be controlled by special ventilation. The mere occupation of a crowded room by human beings does not, however, produce a dangerous concentration of any poisonous substances. Carbon dioxide, under the most extreme conditions which exist in any room above ground never reaches a harmful concentration, and the old idea that specific poisons were exhaled by the human body has proved to be without foundation.

The real problem of ventilation, as Professor F. S. Lee has phrased it, is physical, not chemical, cutaneous not respiratory. The discomfort, the inefficiency, and the serious damage to health which result from bad ventilation are due to overheating, excess or deficiency of moisture and lack of air movement which interfere with the normal elimination of heat and moisture from the surfaces of the body. The victims of the Black Hole of Calcutta died of heat stroke, not of oxygen starvation. The primary essen-

330 CONTRIBUTIONS OF SCIENCE TO RELIGION

tial in ventilation is therefore the maintenance of a moderate temperature with moderate humidity and moderate and variable movement of the air; and the researches of the New York State Commission on Ventilation have recently indicated that the provision of such conditions is of the very first importance from the standpoint of health, comfort and efficiency.

The first step in securing proper air-conditioning is therefore the regulation of artificial heat sources, since overheating is often due to neglect of this factor rather than to any lack of ventilation. The thermometer should be an essential article of furniture in every schoolroom and workroom, and the passing of 68° in the winter season a signal for remedial action. Where there are many persons in a room the provision of some special means for removing the air warmed by the bodies of the occupants and providing fresh air, cool but not too cold, to take its place, will be essential. This is the modern conception of ventilation. Windows alone may serve for this purpose where crowding is not great, as in the wards of a hospital. In the school room, the studies of the New York State Commission on Ventilation have indicated that excellent results may be secured by the use of window inlets with slanting window boards to deflect the air upward and radiators below to temper the incoming air, and with gravity exhaust ducts to remove the vitiated air from near the ceiling. In auditoria, and in many factory workrooms special fan ventilation with forced draft will be necessary.

It should be pointed out that the adjustment between the body and its surrounding atmosphere is influenced by clothing as well as by ventilation, so that the hygiene of clothing is an integral part of the problem here discussed.

In special industrial processes the health of the workers may be menaced, not only by the toxic fumes to which reference has been made above, but also by poisonous or irritant dusts. Stone workers, grinders and polishers, sand-blasters and other workers exposed to hard silica dusts may show a tuberculosis death-rate ten times the normal as an effect of the predisposing effect of these crystalline particles. In all such cases special provision must be made for the control of such hazardous atmospheric conditions, by the use of exhaust fans or masks and respirators.

THE ACCIDENT HAZARD

An adequate supply of food and water, and a suitable atmospheric environment are the three positive conditions of human life. We may now pass to the first of the negative conditions, protection against mechanical violence.

The accident hazard first attracted general attention in this country some fifteen years ago in connection with the hazards of industry. Nearly ten thousand persons are killed every year on the railroads of the United States, between two and three thousand in the mines, and an unknown number, probably in the neighborhood of fifteen thousand, in other industrial occupations. The rates in the past have been, and in many industries still are, largely in excess of those which obtain in the leading countries of Europe.

Certain far-sighted concerns, such as the International Harvester Company, and the United States Steel Company, began about fifteen years ago a vigorous and concerted campaign to reduce the accident hazard, and the passage by the states of Workmen's Compensation Acts which made the burden of industrial accidents a direct financial charge against each industry furnished a powerful stimulus toward the extension of such activities. Safety engineers were employed to safeguard moving machinery, to supervise the handling of materials, to fence pits and platforms, to improve lighting and to attend to a thousand and one details of factory housekeeping which make for safety. Systematic campaigns were inaugurated for enlisting the essential coöperation of the employees; and medical departments were installed to render the prompt surgical care which so often prevents a minor injury from becoming a serious one. As a result of such concerted efforts the number of accidents and the time lost as a result of accidents have been reduced in many industries to less than a third of the original figures.

More recently, in the main during the past five years, the public has been awakened to the fact that the accident hazard is much more than an industrial problem. As other preventable causes of death have been controlled the relative importance of street accidents, burnings, drownings and the like has become steadily greater; while automobile accidents in particular have risen rapidly, with increasing use of this mode of transportation, to a total figure of some 15,000 deaths in 1922. This problem has recently been attacked with vigor in many cities, notably in Detroit; and the success of Safety Week, as staged in New York, and other cities, has furnished encouraging evidence that street accidents can be controlled by proper regulation and by the education of the general public.

CONTROL OF GERM DISEASES—WASTE DISPOSAL

There remains for consideration the last, and in many respects the most important of the basic conditions of healthy living, protection against the attacks of the living foes of man. Aside from such countries as India, where wild beasts and poisonous serpents still take their toll of human life, this means in general the control of the parasitic diseases to which man is heir.

The work of the last half-century, following on the fundamental lines laid down by Pasteur, has made it clear that the germs (microbes of plant or animal nature), which cause these diseases are like other parasites specially adapted to life in the body fluids of the "host" upon which they prey, in this case of man. They have lost, in the course of evolution, the ability to grow and multiply in water or earth or other parts of the external environment. It is to the human body that we must look, then, for the original source of the infecting agents, or in a few instances, to the body of one of the higher animals such as the cow which may transmit to man the germ of bovine tuberculosis. Every case of communicable disease involves the rather direct transfer to the person affected of infective material from the body of some other individual (or in certain cases, as noted, from the body of an animal). It is the objective of the sanitarian to prevent this transfer from taking place.

There are three chief agencies by which the dissemination of disease germs is actually effected; (a) water and milk and other articles of food and drink, (b) more or less direct personal contact, and (c) insects which play a rôle in the spread of many diseases. These three modes of infection are often grouped for purposes of popular exposition under the convenient alliterative headings— food, fingers and flies.

In general, and aside from diseases transmitted directly by biting insects, the disease germs first leave the body of the infected person in the discharges from the nose and throat or those of the bladder and intestines. The control of the nose and throat diseases is an exceedingly difficult problem which will be discussed in a later paragraph under the Contact-borne Diseases; but the care of the bowel discharges which plays a primary part in the eradication of such diseases as cholera, typhoid fever, and hookworm infection obviously is a problem by itself which deserves brief special consideration before passing on to a consideration of the three chief modes of infection enumerated above.

It is difficult for those who are accustomed to the excellent sanitary provisions which now obtain in large cities, and in prosperous countrysides to realize fully the conditions which existed in New York and London fifty years ago, and which exist to-day in tropical countries, and in large rural sections of our own southern states. The heavy burden which hookworm disease lays upon large areas of the world's surface is due entirely to the lack of ordinary sanitary precautions in the disposition of human excrement. It is fundamental in any scheme of sanitary control that such discharges should be received in tightly inclosed receptacles or vaults protected against the access of flies; and systematic supervision of privies along these lines has yielded remarkable results in the reduction of typhoid and hookworm infection.

From the standpoint of the individual householder the ideal plan is that which involves the prompt removal of all the excretal wastes of the household by a modern system of sewers. Sewerage, however, introduces new problems of municipal sanitation, since at the end of the trunk sewers there must be some provision for the sanitary disposition of the accumulated filth of the community, suspended and dissolved in the large volume of liquid which the water carriage system involves. The discharge of untreated sewage into the nearest watercourse may in some cases be permissible; but we must be certain in any given instance that there is no serious pollution of water supplies, bathing beaches or areas from which shellfish are taken, and also that gross and offensive nuisances are not created by the subsequent decomposition of the organic matters thus discharged.

Hence there has arisen the complex branch of sanitary engineering which deals with the disposal of sewage. We have to-day at our disposal various processes of screening and sedimentation for the separation of the solid constituents of sewage and various processes of treatment (intermittent sand filters, contact beds, trickling beds, activated sludge treatment) by which the organic materials present may be subjected to the action of oxidizing bacteria and, by them, transformed into a harmless mineral form. It is possible by the use of such processes to change city sewage into a bright clear water which would be safe to use for drinking; but it is seldom necessary or wise to carry purification to any such point. An engineeer has been defined as "a man who can do for one dollar what any fool can do for two." The problem of the sanitary engineer is to treat a given sewage, under a given set of local conditions, so as to avoid nuisance and danger to health at a minimum cost.

A word should perhaps be said here in regard to the disposal of the solid wastes of the city—garbage, ashes and rubbish—although this is, as a rule, a problem of æsthetics rather than public health. Here, as in the case of sewage treatment, there are many methods to choose from. Feeding garbage to hogs, incineration and reduction (for the recovery of grease) may all prove satisfactory under proper control; and the main essential in a given case is that the local problem should be comprehensively studied by a competent expert before a plan is adopted.

SANITATION OF WATER-SUPPLIES

Passing now to the agencies which play an important part in the transmission of disease germs from one person to another, the first problem to demand attention is that of public water supply. In the past such supplies have produced some of the most terrible epidemics of cholera and typhoid which are on record; and even to-day the consumption of sewage-polluted water every now and then takes its toll of human life.

A first step in the avoidance of such dangers is of course

the protection of water supplies which are to be used for drinking by the removal or treatment of sewage discharged into them and by the strict control of the use of their shores for bathing, boating, picnicking and the like. Except in remote regions where the population is scanty such protective measures can never, however, be wholly adequate to secure a safe water supply, since the pollution of the waters by a tramp or a passing motorist might at any moment produce the gravest results. In addition to the supervision of the watershed, modern sanitation demands that a more positive safeguard should be provided in the form of some process of purification.

The simplest way to purify water is by storage. In passing through a large lake or reservoir, the disease germs, adapted to the rich warm tissues of the human body, gradually die out and storage of several weeks' duration will render even a polluted water safe. One must be certain, however, that the expected storage actually takes place, for many instances are on record in which pollution has been carried across even a large reservoir by local currents induced by winds, so rapidly as to permit the large scale transmission of disease.

A more certain and reliable procedure is slow sand filtration, which consists in passing the water through some six feet or more of fine sand by which the disease bacteria are held back and retained until they perish. Great basins are built with a layer of properly selected sand and underdrains below for carrying off the filtered water, and they are operated continuously, the water from river or lake standing at a more or less constant level over the sand, constantly flowing in at the top and out from the underdrains below. Plants of this type have proved capable of purifying even such polluted waters as that of the Merrimac River at Lawrence or the Hudson River at Albany and rendering them safe for human use.

If six feet of sand in such a filter proves an effective purifying agency, we should expect that by the natural filtration to which ground waters are subject they would be purified in a similar manner. Such is indeed the case in sandy soils, but in clay or rock formations there may be underground fissures which will permit the subterranean transport of pollution for considerable distances to a well or spring. As a rule, however, the commonest source of pollution of the farmyard well is surface wash entering from the top. A tight well curb raised above the surface of the ground, and a tight well platform will prevent such surface pollution, and will insure the safety of a well water in sandy soils.

The slow sand filter does not operate successfully with waters like those of the Mississippi Basin which are rendered turbid by the presence of large quantities of clay. For such waters we have another process known as rapid mechanical filtration, in which the filtering action of sand is reënforced by adding to the water before treatment a chemical like alum which produces a heavy flocculent precipitate. With a layer of this precipitate on the surface of the sand the fine clay particles will be removed and filtration can also be accomplished at a much more rapid rate. The choice between the two methods of filtration in a given case depends in the main on the amount of clay which may be present in the water.

Finally, it is possible to effect the direct disinfection of water supplies by the use of various chemical and physical agents. The boiling of water (usually precedent to the making of tea), is a universal practice in the Orient and furnishes a barrier of the first importance against the spread of water-borne disease. Ultra-violet light, ozone and other disinfectant agents may be used on a small scale; but by far the most valuable procedure of this kind in the disinfection of water by the use of chlorin gas or bleaching powder. A large majority of the water supplies of the United States are now treated with chlorin. The process is a very economical one, and forms an invaluable adjunct to storage or filtration, putting the final touch on a water which would perhaps be otherwise generally, but not always, safe. With this addition to the methods of purification at our disposal there is no longer excuse for any community which permits its citizens to run the slightest danger of water-borne disease; and the courts have in numerous instances rightly awarded damages to individuals who had suffered from such maladies.

SANITATION OF MILK AND FOODS

Next to water, milk has, in the past, been the food most commonly responsible for the propagation of disease on a large scale. Milk may transmit the germ of bovine tuberculosis from the cow to man. It may serve as the agent for the spread of diseases of human origin like typhoid fever and septic sore throat; and, among infants, milk which is not specifically infected but merely decomposed by the action of various bacteria is a fertile source of summer diarrhea. Here, as in the case of water supply, some degree of safety can be secured by the protection of the milk supply against dirt and decomposition, by farm and dairy inspection, by the tuberculin testing of cattle, and by prompt cooling to check bacterial decomposition. It is practically impossible, however, to secure anything like complete protection by these means. Clean milk is not necessarily safe milk; a carrier may at any time come into contact with the milk in barn or dairy, and if this occurs, a single cough over the pail, or the touch of a finger, may sow the seed for hundreds of cases of disease. Fortunately, again as in the case of water supply, we have simple processes of purification which make it possible to secure a really safe product. This end is best accomplished by pasteurization, which consists in heating the milk to a temperature between 140° and 145° F., and holding it there for at least twenty minutes. This process destroys all disease germs and the vast majority of the ordinary bacteria of decomposition as well, and it effects this result with no deleterious effects upon the milk except the weakening of one of the vitamines. Milk is not a particularly reliable source of this particular vitamine in any case, and where even raw milk is used as the sole food of infants, it should be supplemented by orange juice or tomato juice to supply this substance. It is desirable to keep milk as clean as is practically possible at all stages in its production and handling, but it is essential to safety that it should be finally purified by proper pasteurization. With the growing use of this practice, which is required by Health Department regulation in many cities for all but a very small amount of milk of specially high grade, milk-borne disease has been reduced to relatively small proportions.

Raw shellfish rank next to water and milk as factors in the spread of food-borne disease; but this source of danger, too, is now generally controlled by effective supervision of the areas from which such foods are taken. Almost any article of food or drink may, however, be infected by handling in the kitchen just before it is served. The famous case of ''Typhoid Mary'' is the classic illustration of this possibility. She was a cook by profession, but also a typhoid carrier, perfectly well herself, but cultivating typhoid germs in her body, and more or less constantly distributing them with her discharges, and was the cause of half a dozen small epidemics in the various households where she was employed and later of a large epidemic in a hospital. A maximum of cleanliness in food preparation processes, and the periodical examination of food handlers for the detection of carriers are the best safeguards against infection of this type.

Food poisoning, often erroneously called ptomaine poisoning, is generally due to the infection of foods with specific bacteria derived either from a carrier in the kitchen or from the flesh of a diseased animal; or, in the case of the particular form of food poisoning called botulism, to the presence of a type of bacterium which grows and produces a powerful poison in improperly handled canned goods.

Finally a word should be said as to the supposed danger from added harmful chemical ingredients in foods, adulterants, preservatives, coloring matters and the like. Injury from such substances is very rare indeed, and from a practical standpoint the menace to be feared is a menace to the pocketbook rather than the health. The Pure Food Laws are more than justified in so far as they guarantee that the purchaser shall obtain the thing he is paying for. From a health standpoint, however, it is the uninstructed housewife, rather than the wicked grocer, who must be dealt with; and the teaching to the public of the principles of dietary hygiene, as discussed in an earlier paragraph, is the most important factor in promoting a soundly nourished population.

CONTROL OF THE CONTACT-BORNE INFECTIONS

The second group of diseases, those which are spread by more or less direct contact between one human being and another, are much more difficult to control than the infections disseminated by food or drink. The chief contact-borne diseases are those in which the discharges from the nose and throat play a primary part, such as pneumonia, influenza, diphtheria, whooping-cough, scarlet fever, measles, smallpox, infant paralysis, epidemic cerebrospinal meningitis, and the common cold. In most of these diseases well carriers are common, and the frequency with which the fingers or objects handled by the fingers go to the mouth or nose, makes it extraordinarily easy, with the aid of pencils and coins, door knobs and telephone receivers, push buttons and trolley straps, for the exchange of infected saliva to occur.

The efforts of the health officer are directed, first of all, toward obtaining the reporting of cases of communicable disease, and then securing the isolation of each case under such conditions as shall minimize the danger of transfer to others. The patient should be cared for in a separate room by an attendant who makes certain to disinfect everything that has been soiled by the discharges of the patient before it leaves the sick room. Such daily and hourly precautions alone will suffice to control the spread of infection. If they have been taken during the course of the disease, fumigation at its close will be unnecessary; if they have not been taken, fumigation will be useless, because relatives and friends will long before have been infected. Fumigation after the close of the isolation period is therefore no longer provided by the most progressive health departments; but instead a rigorous care is exercised in securing proper isolation during the course of the disease.

Unfortunately, even the cases of disease which are reported to the health department and effectively isolated are often reported so late that many people have already been exposed and the seed sown for a crop of secondary cases. Measles, in particular, is most contagious in its very early stages when it resembles an ordinary cold. It is therefore of the first importance that "contacts," as they are called, persons who are known to have been exposed by contact with infected individuals, should be either isolated or kept under close observation until it is certain they are not likely to develop the disease in their turn. In diphtheria and cerebrospinal meningitis such contacts can be examined by bacteriological methods to determine the presence or absence of the specific organism. In other diseases like scarlet fever or measles the contacts are kept under supervision for a fixed time, corresponding to the "incubation period" of the disease in question, the period which normally elapses between the time when the germ of that particular disease enters the body, and the time when the first definite symptoms make their appearance.

For attaining a maximum of results in the attempt to control the contact-borne infections we must rely, as in so many other instances, very largely upon education. Since no system can possibly insure the detection of all carriers, we must strive to develop what has been called the "aseptic sense," an instinct which will keep away from mouth and nose everything which is not bacteriologically clean; and we must attempt also to cultivate "the sanitary conscience" which leads those who are infected with any communicable disease, even if it may seem on the surface to be only a common cold, to keep as far as practicable out of contact with others, and never to shake hands or handle things that others will be liable to touch. Above all, we must emphasize the importance of keeping infected people away from young infants. Measles and whooping-cough cause far more deaths to-day than scarlet fever and typhoid; and in the overwhelming majority of cases their victims are infants and young children. Every year, every month, for which attacks of such diseases can be postponed, renders the danger of a serious result by that much more remote.

THE ARTIFICIAL CONTROL OF IMMUNITY

We possess another and quite different set of weapons against many of the communicable diseases discussed above —the sera and vaccines by whose use it is possible to build up a special and powerful vital resistance against specific diseases, either before or after the germs in question have invaded the body. It is perhaps permissible to include a brief discussion of the resources of vaccine and serum therapy and prophylaxis under the topic of Sanitation, since the use of these substances forms an important part of the official public health program.

The recovery from an infectious disease is probably always accompanied by the development of a greater or less specific immunity against the causative agent of the disease. The principle involved in the use of vaccines is that the injection into the body of a very few virulent organisms, of a larger number of weakened organisms, of a suspension of killed organisms, or of the extract from them, may produce in the body a reaction which leads to a specific immunity of a similar kind. Jenner's discovery of vaccination as a preventive of smallpox furnished the first brilliantly successful example of such a procedure, a century and a quarter ago; and the studies of Pasteur established the principle on a broad basis of experimental evidence, and opened the way to its application in a wide range of other diseases. To-day we have not only the vaccine of Jenner, which has completely eradicated smallpox wherever it has been conscientiously applied; we have also a vaccine against typhoid which produced the most brilliant results in the protection of our armies during the Great War; the anti-rabic virus of Pasteur which is essentially a vaccine, protecting those who have been bitten by a mad dog against the development of rabies; vaccines of considerable, though not complete effectiveness against cholera and plague, and many others of less general importance. The vaccines against smallpox and typhoid fever are of such demonstrated and general efficacy that every possible means should be taken to foster their universal use.

Such vaccines as those described above are used to develop active immunity in the treated individual; their effect is more or less lasting, and they are commonly used for protection against anticipated infection in the future. In many diseases it is possible to use similar vaccines on an animal, to draw off some of the blood serum of the animal containing the active principles of immunity which have been produced by its reactions, and to use this serum, containing these principles, for producing a temporary passive immunity in human beings. Such an immune serum is generally utilized as a therapeutic agent in the treatment of a disease actually in process, since the passive immunity produced is powerful, but transient in its effects. Diphtheria antitoxin is the outstanding example of a serum of this kind, since it effects an almost certain cure if applied early enough in the course of the disease. There is a serum for epidemic cerebrospinal meningitis which has cut down the mortality from this disease to a small fraction of its normal figure; there is a valuable serum for the treatment of tetanus, a serum of great promise for the treatment of certain forms of pneumonia, and many more. Rapid progress is being made every year in the preparation of new sera and vaccines for other diseases, and it is altogether likely that the greatest progress in the field of public health during the next decade will be achieved along this line.

It should be noted in passing that in the case of diphtheria our armamentarium is already remarkably complete, and that the considerable mortality from this disease which still persists constitutes a serious reproach to all of us. We can recognize cases and carriers of diphtheria by a simple culture method. We can cure an infected person by the use of antitoxin. We can determine whether an individual is susceptible to the disease or is naturally immune to it by the use of a test called the Schick test; and we can produce a lasting active immunity against it by injections of a mixture of toxin and antitoxin which acts essentially as a vaccine. If as a community we were alive to these possibilities we should give toxin-anti-toxin treatment to all infants at about one year of age, as we give smallpox vaccination now; and diphtheria, like smallpox, in well vaccinated communities, would become a medical curiosity.

THE INSECT-BORNE DISEASES

Finally, there remains for consideration the group of diseases in which the transfer of the specific germ is effected by the agency of an insect host, a group of tremendous importance in earlier days and still of major significance in tropical regions.

The relation of an insect to the transmission of a particular disease may in some instances be purely incidental and occasional, while other infections are spread solely through the medium of their insect hosts. The part played by the ordinary house-fly in the spread of typhoid fever and other intestinal maladies is an example of the first relationship. Typhoid is not generally spread by flies, but where human excrement is carelessly handled and flies are abundant (as in the American army camps during the Spanish War) the fly may play a significant rôle in the dissemination of this disease. It is of real sanitary importance, therefore, that fly-breeding should be controlled by keeping the neighborhood of dwellings free from waste materials, and particularly by the proper care of horse manure which is the favorite breeding place of this insect.

A more direct relation to an insect host is manifested by bubonic plague, the terrible Black Death of the Middle Ages, which is said to have carried away one fourth of the population of Europe. Under certain circumstances this disease may assume the pneumonic form in which it is spread from person to person by direct contact; but the typical bubonic plague is a more or less chronic disease of rats and other rodents which is transferred, first from rat to rat, and then from rat to man, and from man to man by the flea. Its practical control can best be effected by a systematic campaign of extermination directed against the rat. After slumbering for some centuries this disease broke out again in Asia in 1871. It killed six million people in India between 1896 and 1906, and has since spread to seaports in every part of the known world; but everywhere, since its mode of spread was discovered during the Indian outbreak, it has been promptly checked by the methods of rat control.

Another disease, of sinister significance in medieval times, was typhus fever, often called camp fever and jail fever, which decimated the troops of Napoleon and dogged the footsteps of many a military leader before and after. The knowledge that this infection is spread by the bite of the body louse made it relatively simple to control it during the Great War by the enforcement of rigorous and systematic cleansing of bodies and clothing. When the machinery of civilization broke down in Revolutionary Russia it ravaged that unhappy country; but the admirably effective sanitary measures enforced by the Polish Government, aided by the Health Section of the League of Nations, have protected Europe from its extension westward.

Our major problem of insect-borne disease in the United States is malaria, which Sir William Osler once estimated to be of all infections the one which lays the heaviest burden of sickness and disability upon the human race. Since the discovery, twenty-five years ago, that malaria is spread by the bite of a particular type of mosquito it has been demonstrated again and again that the incidence of this disease may be reduced to negligible proportions by the control of mosquito-breeding (effected through ditching, oiling or stocking waters with fish which devour the mosquito larvæ), or by thorough screening of houses to keep out the deadly insects.

Another mosquito-borne disease, yellow fever, furnishes an even more brilliant example of the achievements of sanitation. The mode of spread of this scourge of the Tropics was discovered in 1900 at Havana by a group of American Army surgeons, Reed, Carroll, Lazear and Agramonte. It was necessary to experiment with human subjects, and the officers in charge began with themselves, Carroll suffering a severe attack and Lazear giving his life in the cause. Yet volunteers for the grim task were never lacking; and the heroism here displayed brought swift results. In six months the particular mosquito responsible was discovered; and the next year the new knowledge was applied so promptly that there were but eighteen deaths from yellow fever in Havana in 1901 and none in 1902, as against 750 a year for the preceding period for which we have records. It was this discovery at Havana which a few years later made possible the building of a canal across the hitherto plague-ridden Isthmus of Panama; and to-day the International Health Board has taken the offensive in a world-wide campaign which is eradicating this disease year by year from one country after another and which promises within a decade to eliminate yellow fever forever from the roll of Azrael's ministers.

ACHIEVEMENTS OF SANITATION, AND THE PROMISE OF THE FUTURE

The record of the achievements of sanitary science in the past quarter-century, taken as a whole, constitutes one of the most significant chapters in the history of civilization. The death-rate of New York City for example has been reduced from 25 deaths per year per 1,000 persons in the population in 1890 to 13 per 1,000 in 1920, an achievement which involves the saving of 183 lives every twenty-four

348 CONTRIBUTIONS OF SCIENCE TO RELIGION

hours. For the United States as a whole the reduction in the death-rate from four diseases alone (typhoid fever, diphtheria, infant diarrhea and tuberculosis) between 1900 and 1920 corresponds to a saving of 230,000 lives a year. For the state of Massachusetts, where statistics are available for a longer period, the average length of human life was 40 years in 1855 and 55 years in 1920, a prolongation of 15 years. Think what it would mean if on the deathbed of a given individual a reprieve of fifteen years could be granted; yet that is just what has happened, on the average, to every man, woman and child in the community.

And the end is not yet. A recent estimate by Dr. L. I. Dublin, statistician of the Metropolitan Life Insurance Company, indicates the possibility of adding another ten years to the average span of human life. Sanitation, in the narrower sense cannot, however, alone and unaided, accomplish the tasks of the present and the future. The environmental plagues and pestilences have been largely conquered. A victory over the subtler dangers of maternal and infant mortality, of tuberculosis, of cancer, of the degenerative diseases of later life, requires an expansion of the public health program to include such an organization of the medical forces of the community as shall make the physician a true agent of prevention and not merely an alleviator of architecturally completed disease. To bring about such a reorganization, in ways that shall preserve the freedom and the initiative and the high standards of the medical profession, is the major health problem which now lies before us.

PART III

RELIGION, THE PERSONAL
ADJUSTMENT TO ENVIRONMENT

CHAPTER XV

THE EVOLUTION OF RELIGION

By SHAILER MATHEWS [1]

THE use of the term evolution in connection with religion is subject to at least two objections. On the one side are those who insist that religion is the gift of God, and therefore has no historical development. And on the other the biologist may object to the use of the term in any such general sense as a student of social science must adopt.

To the first critic it may be replied that, when he asserts or implies that religion has not developed like other elements in human experience, the facts are against him. Whatever may have been its origin, religion exhibits phenomena akin to those observable in social institutions to which the term "evolution" may legitimately be applied. The old distinction of the Deists between the natural and a revealed religion has been outgrown, not so much because it did not involve large elements of truth, but because as a final answer to the problems set by the history of Christianity it failed to take into account those psychological and sociological factors with which the modern student is particularly concerned. All religions are phases of religion.

To the other class of critics it must be replied that if biologists ever had a monopoly on the term evolution their exclusive rights have long since expired. The conception given the word by the "Origin of Species," and general biological usage is a particular phase of a view of the world

---

[1] Professor of Historical Theology, University of Chicago.

as old as reflective thought. The service which biology has rendered the social sciences at this point has largely been confined to the region of method, vocabularies, and analogies. If these analogies have too often been over-emphasized and made to do yeoman service in the name of some non-biological science, they have none the less made it possible to realize that, whatever precise definition may be given the term evolution, there is a large measure of similarity between certain processes in social history, and certain others in the building up of cellular organisms. Outside of the strictly biological sciences the word must be used in a large sense, but it is not identical with mere change or growth. It is possible to trace religion as one of the functional expressions of life itself through increasingly complicated and more highly differentiated activities and institutions, as that life both of individuals and societies seeks genetically to adjust itself more effectively to its environment. The result of such vital activity is to produce, as it were, species of religions, between which, as for example Brahmanism and Mohammedanism, there is only a generic likeness.

To justify the legitimacy of the use of the term evolution in a reasonably strict sense, this chapter will discuss (1) what religion is; (2) its development into species of religions according as its expression has been conditioned by its environment; (3) the persistence of vestiges of lower religious forms, concepts, and institutions in the more highly developed; and, (4) the struggle for the survival of the socially fittest among religions.

THE NATURE OF RELIGION

There have been times in which men have endeavored to arrive at the conception of religion by abstracting from Christianity its characteristic elements. Other attempts have been made to extend this process of abstraction to all religions, and thus to discover that which is, to so speak, a generic concept. The difficulty with such search after a bit of scholastic realism is evident. Generic religion never existed apart from religions, and religions never existed except as interests and institutions of people. There is imperative need that all students of the subject, and especially theologians, should emancipate themselves from scholastic abstractions, and frankly recognize that religion is not a thing in itself, possessed of independent, abstract, or metaphysical existence, but is a name for one phase of concrete human activity. It is only from a strictly social point of view that either religion or religions will in any measure be properly understood. We know only people who worship in various ways and with various conceptions of what or whom they worship.

Yet while men possess *religions* and not merely *religion*— religions of all sorts, from the simplest custom of the savage to the profundity of Brahmanism and the redemptive gospels of the Buddhist and the Christian—the comparative study of human activities expressed in these different religions has, however, discovered within them *religion* as a common divisor, as it were; viz., a particular functioning of life itself, as truly and universally human as the impulse of sex and self-preservation.

If we attempt to formulate this common element, or rather to describe this functional expression of life expressed in all religions, we must study comparatively both the highly developed religious systems and the simplest type of religion as it exists among primitive people. That is to say, while not overlooking the more complex systems as a means, so to speak, of determining the direction taken by evolution and thus better fitting ourselves to appreciate religion as never absolutely static, we must study the simplest religious organisms in order to understand the more complicated. To push the biological analogy farther, it might be said that the "cell" of religion is man's conscious attempt to place himself in help-gaining relationship with those superhuman forces in his world upon which he realizes his dependence, and which he treats as he would treat persons whom he wished to aid him.

It is obvious that the content of such a formal definition will vary according to the conception of what constitutes this superhuman environment; and that this variety of estimate will affect the methods which a man adopts in making that environment propitious. A study of even the most primitive religion leads one to two convictions apparently paradoxical: religion does not necessarily imply a belief in a supreme person, and yet, in religion, environment is conceived of in the same way that men conceive of persons. Therein the functioning of life in religion differs from the functioning of life in the satisfaction of the impulse of sex and food-seeking. True religion does not, as Monier-Williams would insist, postulate the existence of one living and true God of infinite power, wisdom, and love. That would exclude too many religious customs and rites. Men have worshiped fetishes or animals or sacred stones. Such objects are regarded as elements in the environment which affect human interests, and, therefore, without being of necessity consciously personified, are treated as if they were personal.

There are a number of theories undertaking to show how this attitude of mind was induced; but all are more or less unsatisfactory. Some find the cause in fear, or dreams, or regard for ancestors, or the appetencies of sex. Doubtless there is truth in all of these hypotheses, but we are not absolutely sure as to just how religion came into existence any more than we are sure how human life itself arose. We can, however, see clearly that the functional significance of religion is an elemental expression of the second of the two elemental impulses of life itself, namely,

to propagate and protect itself. Religion is life functioning in the interest of self-protection. It differs from similar functional expressions of life in that (1) it treats certain elements of its environment personally (though not
necessarily as a person), by utilizing social practices and
ideas as forms of worship, or as patterns for beliefs, and
(2) it seeks to make these elements friendly and so helpful.
One or the other of these two elements has almost invariably been overlooked in studies of religion, but both are
indispensable to the concept. Religion utilizes personal
experience and uncompromisingly presupposes personalism not, let it be repeated, in a sense of any systematic
world-view, but, in a sense that doubtless unconsciously at
the first, but with ever-increasing clearness of conception, it
treats the environment as it would treat human beings;
and religion is just as uncompromisingly functional, not
only in adjusting the individual or the group to its environment, but also in the attempt to adjust the environment
personally considered to the person or the community.
Thus Schleiermacher's conception of religion as a feeling
of dependence is only part of the truth. To it must be
added the conscious effort after reconciliation. It is this
two-fold modification of the elemental functioning of life
in the interest of self-preservation that distinguishes religion from so many activities with which it has been intimately associated, like hunting and grain-planting, marriage and burial.

Obviously the inception of this radically human attitude
toward its world is lost in the unrecorded struggles by
which humanity raised itself above other forms of animal
life with which it is genetically united. But one's ignorance here does not impugn the fact that such a use of
experience was actually made.

Sometime, somewhere—just when and where it matters
not—there appeared a man who first of all living creatures,

with the new impulses of a genuine person, attempted to adjust himself consciously to the outer world upon which he saw himself dependent, by an attempt to make that outer world favorable to himself. It makes little difference how he conceived that outer world or which one of its particular aspects first impressed him. Any one of the various theories of the origin of religion might here suffice. The essential thing is that, in his passion to protect his life and to insure his continuous existence as a person, he attempted consciously to enjoy or to win the favor of the extra-human environment with which he found himself involved. And that, so far as we know, no animal other than man ever attempted to accomplish.

Nor is it necessary to insist that all religions are genetically related in a sense that one has been derived from another. The historico-religious method at the present time is in danger of mistaking similarities between religions for genealogical relations. Thus in the comparative study, let us say, of Christianity, there is strong temptation to insist that elements of Babylonian myths go to constitute the very content of Christianity. That a certain degree of genealogical relationship in this particular case may have existed may well be admitted, but a too rigorous application of the comparative genealogical method in the study of religion is certain to distort the facts. If there is anything undeniable in the study of society it is that human nature is essentially the same, and that when facing the same social needs it functions in a generic sort of way. A striking illustration of the fact that independent activity of individuals produces similar results is to be seen in a study of inventions. The commonest occurrence is for men subject to the stimulation of similar social need, in absolute independence of each other, to produce instruments and processes practically identical. An even more striking illustration of this general truth is that all civilizations precipitate practically the same moral codes when they arrive at the same stage of complicated social life. So in the case of religions, the striking similarities which occur between religions belonging to the primitive class and religions belonging to the highly social class are not to be interpreted as necessarily involving imitative, or in fact any, historical relationship. Such similarities both in institution, and in process of evolution, can often be sufficiently well accounted for by a generic religious impulse in humanity which tends to produce customs, rites, institutions, and creeds in answer to individual and social needs.

The evolution of religion as an attitude viewed historically, is nothing more or less than the organization of religions by the differentiation, through the use of social experience, of the practices, institutions, philosophies, by which men have attempted to justify, rationalize, direct, and give value to this phase of the elemental impulse of personality.

The Evolution of the Personal Interpretation of Environment

It will be understood from what has already been said that the term extra- or superhuman environment does not always necessarily involve personality. What the term means is simply some power other and more than human which a man regards as having influence upon his life and fortunes. The fact that such elements of the environment *are treated* as if they were personal, is only to say that religion involves an extension of experience over into environment as a means of interpreting that environment in the interests of a helpful reconciliation. Such an act is not unlike the way in which, to speak figuratively, a living organism makes the assumption that its environment discovered by experience is capable of forming a part of a dynamic situation. Thus far Ward is correct in saying that religion is in man what instinct is in animals. But only in so far, for did an animal ever seek to placate nature?

The essential matter in the evolution of religion as in all evolution is the transformation of the original organism through its relation with its environment and the nucleating about itself—if the figure may be allowed—of the cells of other experiences into species of the same genus. And this is accomplished by the transformation of the mass of experience with which humanity adjusts itself to its environment to which it must submit and from which it must derive assistance.

1. Primitive religions generally deal with environment directly. The primitive gods in the earliest strata of survivals and literature in which we can trace religious concepts were natural forces. The heavens and earth, fire, water, and wind, the sun, moon, and planets—these natural objects were worshiped but they were not personified. Man found himself face to face with the awfulness of Nature. He saw how dependent he was upon this nature; how the rising of the river would flood and sweep away his hut; how the rain would come from heaven to give him grass for his cattle, how the sun would drive the animals he hunted into the deep forests. He naturally wanted to make the river and the heavens propitious. A little later, he very likely turned animistic and regarded natural forces as the home or the visible expression of personal gods; but at the start he worshiped unpersonified natural objects. But he began to treat them personally—as he treated the other members of his tribe or other tribes when he wanted help.

If we go even farther back than philology can carry us and study religion as we discover it in the most primitive folk, we find corroboration for this view, although with this difference: there seem to be some tribes that have not risen to the conception of the great natural forces as those that are to be appeased, and who therefore concern themselves rather with items in their natural environment. In fact, anything unusual is apt to be regarded by primitive men as a good or a malign influence. In either case it needs to be treated with respect, and if possible placated. A rock over which some one has fallen, a cave in the darkness of which some one has been lost, a curious root that was discovered when one became ill, a tree that had been struck by lightning—all have been regarded as operative forces in a man's situation which needed in some way to be placated.

Here, too, an early step was to regard these natural objects as the residence of some spirit, good or evil. Thus fetishism arose as a sort of limitation of the lesser nature-worship. Not all natural objects were significant, and even those which were might lose their meaning if the spirit abandoned them.

It is difficult to draw a distinction between magic and religion even when religion begins to take on its more social form. Not a few group actions resemble practices usually called magical when followed by individuals. But the witch is different from the priest if for no other reason than that her arts are anti-social. But at least some magic may very likely be treated as the vestige of a rudimentary religion preserved and observed by specially empowered persons. For there is in such magic, e.g., rain-making, that "will to conciliate" as well as to control which, as the complement to the "will to power," is the very sign manual of religion. But this is not to say that religion developed from magic. The fundamental difference between magic and religion lies not in that magic was originally anti-social and so nefarious, but that in the course of social evolution it is seen to be so. As religion develops, certain rites are seen to apply direct compulsion to superhuman powers, or the impersonal principle that like affects like through the agency of a specially empowered person. Naturally such rites are but religion seeking to conciliate superhuman influences by means implying personal relations and attributes. This distinction between personal and impersonal is gained through the increased social experience. That practice, which once implied a certain personal analogy, is seen to be irrational and so impersonal, and ultimately anti-social. The primitive religion, thus outgrown, becomes magic, and although socially condemned, continues as a survival. And the reason why it is condemned is in large measure the development of a knowledge of natural processes. A growing science thus relegates certain elements of a religion to superstition.

Similarly, too, in the case of the worship of dead ancestors, a stage in religious development is to be found all but universally in simple civilizations. Whatever may have been the origin of such a custom it is sufficiently clear that the dead were regarded as important factors in determining good and evil fortune. To propitiate them is therefore good policy as well as tribal piety.

2. With the emergence of actual tribal organization a new phase in this religious interest appeared. A developing civilization does not always, it is true, immediately react upon the conception of the god, but in so far as the religious concept develops, it invariably passes through a stage in which these forces which have been treated *like* persons are treated *as* persons. That is to say, contemporaneously with the development of the clan, religion entered into the stage of naïve anthropomorphic or anthropopathic religions. Such a development was inevitable for people sufficiently constructive to become a part of the main current of civilization. All others, like the Black Fellows of Australia, preserve the religious ideas in forms as primitive as their civilizations. Such personification,

however, does not seem to have proceeded uniformly. In some cases a tribe would have as its own a god who was the personification of some natural force, and would worship him by attributing to him those qualities which, thanks to its social development, the tribe as a whole believed to be the most ideal. Without exception these tribal gods are regarded as normally in a state of reconciliation with the tribe. Generally they are regarded as the fathers of their tribes. In other words, they are believed to partake of the same elemental quality as primitive civilization itself. They are, however, subject to paroxysms of anger evidenced by the defeat of the tribe in battle, by the outbreak of disease, and by various other misfortunes. In such cases they must be placated by gifts. In this we see one of the various contributing influences that made sacrifice a social institution, although there are other influences quite as powerful. At other times a god appears to be particularly favorable in that he sends good weather and good fortunes. At such times his kindness needs to be appreciated by gifts. Thus arises the sort of sacrifice which is not intended to appease but to thank the tribal god for his help.

But the most essential element in the tribal religion is the conception of the god as the supreme member of the tribe. It is true he is not believed to appear frequently, but at critical moments some member is likely to see him and get some word of encouragement or warning. Further, there have been few peoples who have attained the tribal form of society in which there is not some particular person or family regarded as in some way the god's special representative. Such persons instruct the tribe as to the will of the god, serve as priests, and, under the god's direction, establish great feasts of which the god partakes. Probably at this point we find the most important contributing source of sacrifice. The social group includes the god and he shares in the experiences of the tribe, be they sad or joyous. And it should be noted that the rites of religions had their origin in the enjoyment of life as truly as in its misery and fear. Men thought of the gods as their companions as truly as their judges.

This tribal god in some tribes may, so to speak, be assisted by a number of secondary gods; but polytheism is not necessarily an element of tribal religion, and even when a tribe worships several gods it is likely to have one particularly its own. In fact, as the tribal civilization develops, in many cases, particularly among the Semites and the Aryans, it would seem as if there were two classes of gods—those which represent the material forces more or less personified and constitute a sort of superdivine body of deities to whom worship is to be paid as the final sources of good fortune, and, along with these, so to speak, the working class among the gods. Other tribes carry along with their single tribal god a phase of magic which may be said to be the survival of some more primitive religious practice. Similarly, customs, the meaning of which has long been forgotten, may be carried along as essential elements of a developing religion. So important may these customs become as to give almost its full content to the religion.

3. The fact that the tribal god was regarded as, so to speak, the responsible party in tribal history, led to another phase of religion, the monarchical. Such a term is at best unsatisfactory but serves to indicate how the thought of God develops by the extension to him of new political conceptions. The national god must be superior to the tribal chieftain. As a chieftain developed in power by conquest so as to extend the power of the tribe over other tribes, it has been all but uniformly true that the tribal god was regarded as victorious over the gods of the conquered tribes. Thus, as the tribe itself through conquest became

the head of a quasi-nation, did the god become a conquering monarch. Only it did not at all follow that the tribe which had been absorbed or conquered would give up its god. It might continue to worship him in the hope that ultimately he would assert himself, and give deliverance to his people. Or, on the other hand, as the tribe was incorporated into a new political entity, its god might become a member of the royal court of the supreme God. There is many a nation whose religious history shows the struggle between the worship of the two sets of deities. Thus we find, in the history of Israel, a long succession of struggles between the worship of Jahweh and that of the Baalim and the Syrian gods of the high places belonging to the conquered Canaanites. This struggle is likely to be particularly violent when the two sets of gods are brought together, not by war or conquest, but by the intermingling of civilizations.

For conquest is not the only source of the development of the king god. Political development as such leads to this more developed conception. It may often be that a number of tribes have the same god. These may federate, as in the tribes of Israel, religion being the sole or at least the chief bond of the political unity. But even such federation is not necessary for the development of the idea of God. The transformation of the tribe from nomadic to agricultural life has been accompanied by a transformation of the conception of God and has given him new attributes, as in Zoroastrianism. Sometimes this addition has been made through the religious teachers or the priests; sometimes it has been unconsciously due to the rise of new economic conceptions born of social evolution. As the agricultural stage of social evolution has passed into the commercial and urban, the new powers of the chieftains have been used as media for shaping new prerogatives for the god. His relations become less those of the father of the family, and more those of the king, increasingly political and forensic. It is not too much to say that in the case of all tribes whose development we can trace across the various stages of social evolution, the idea of monarchy, which, however different its social institutions may have been, has characterized some period of every developed society, has also colored religions. The god is not subject to the will of the people; the people and their material environment are to obey him. Obedience to his law becomes thus a condition of his rendering his people aid.

At this point certain great religions have made two important transitions:

1. The superhuman monarch of the tribe has come to be regarded as the superhuman monarch of the world, the king of creation. It has not followed that all the other gods have been regarded as nonexistent, for in many cases they have been treated as devils or saints. But the passage to genuine monotheism cannot infrequently be traced through this monarchical stage.

The divine monarch is supreme over human subjects. He arranges nature. The thunder is his voice, the wind his messenger, the earthquake the creature of his will. Men begin to think of him philosophically, and so transcendental may the thought of him become that the effort to realize the now supreme and increasingly ethical conception of his character gives rise to a genuine if naïve theology.

2. The second transition has been the moral elevation of the idea of God. This change has been the work of the prophet. In primitive religion the prophet in any true sense of the word is unknown. There are only medicinemen, necromancers, witches, and the like. But few peoples ever come to the universal monarchy conception of its god without seeing in him the standard of morality. If such a transition is impossible a new god is adopted as the new

conscience needs a more sensitively moral God. If, as in the case of classical mythology, gods are past reformation, they are pensioned off with conventional honors and allowed to pass into innocuous desuetude on some mountain where their example will not injure the morals of young people. In the extent of this moral idealism of its idea of God, the Hebrew religion is unique. It seems to have passed through the earlier stages of religious evolution, but as in no other religion did this eventuate in a monarch of absolute righteousness, hating iniquity. That this was the case was due to the work of the prophets who, from an exceptional religious experience, taught an unwilling nation ideals that were to serve as the basis of the non-monarchical ethical religion of Jesus.

This monarchical conception has given rise to the most precise theologies. It is easy to see why. Political experience is so universal, political institutions are so subject to legal adjustment, and legal analogies are so intelligible, that it has been comparatively easy to systematize religious relations under the general rubrics of statecraft. Thus righteousness has been thought of as the observance of the laws of the god, given through divinely-inspired teachers, and punishment has been attached to the violation of such laws in precisely the same way as to the violation of laws of the king. The pardoning of sins has been a royal prerogative, although sometimes needing justification in the way of vicarious suffering by some competent sacrificial animal or person while the rewards of the righteous have been pictured by figures drawn from the triumphs of earthly kings, just as in primitive societies the future was regarded as the "happy hunting ground."

3. Only a few religions have as yet progressed beyond the monarchical stage. In Brahmanism, religion has been denied content and direction by an impersonal cosmic philosophy, and two of the three great religions of Semitic

origin—Judaism and Christianity—have moved over into a quasi-transcendental personal sphere. But the theologies of even these religions have been developed on the monarchical analogy. In Christianity, however, the influence of Jesus has resulted in the retranslation of the divine king into the divine Father. His own experience here furnished the interpretative analogy. Unaffected by philosophy he expressed religion in terms of most generic experience, and thus may fairly be said to have closed the cycle of purely religious anthropomorphic formulas. But the Christian religion has not been content. It has sought rationalizing formulas or patterns in which to synthesize itself with such elements of its environment as are contained in a growing world-view. Nor is this synthesis the mere establishment of a static situation. All three elements—the world-view, Christianity, and the situation itself—are in process of evolution. Paternity can never serve as a synthetic theological and philosophical concept. True as it is for experience it has been too obviously an analogy for theology. Historical orthodoxy is built on divine sovereignty, but there have already begun to appear signs that in Christianity the social mind is describing that environment upon which men find themselves dependent in terms more consistent with scientific thought than are those derived from monarchy. Herein may be said to be the real crisis in which theology exists in highly civilized countries. Convinced as are men of scientific temperament that the monarchical conception, and an incomplete democracy are inadequate to express cosmic relations, a rapidly developing scientific thought has not yet reached sufficiently distinct conclusions to enable one to forecast exactly the next stage in the evolution of those conceptions by which modern men shall make intelligible to themselves the significance of the religious life.

There are those who insist that there is no next stage;

that the situation in which religion and science find themselves is capable of no further progress; that the future is to be religionless; that humanity is to replace God, and that ethics is to replace religion as the means by which to regulate the impulse toward reconciliation with a personal environment. But this forecast seems to me untenable. Tendencies have developed so rapidly within the past four or five years looking to the justification, from the point of both psychology and sociology, of religion as a normal attitude, that it is hardly likely that the impulse to adjust oneself to the non-human, cosmic environment, conceived of in some personal way, will disappear. We face, it is true, the question as to what is meant by the term "personality" when applied to that appalling environment which astronomy, biology, and geology have discovered and are discovering. In a certain sense we are back again where religion began its evolution. We can no longer think of God in the way of a naïve anthropomorphism. We no longer think of God as sending plagues; we have fastened that indictment upon bacteria. We no longer believe that eclipses are punishments for our sins or that famines and earthquakes are due to divine displeasure. Like our primitive ancestors we are face to face with the forces of Nature. Indeed, we are not altogether sure that we ought to speak even about forces. We are really face to face with the Whole.

THE PERSISTENCE OF SURVIVALS IN RELIGION

But evolution in religion no more than in a living organism is a matter of ungenetic change. Each new stage in its expression perpetuates in a greater or less degree vestiges of previous stages. Religions have their embryology as truly as their physiology. Just as the human body in its present condition has within it the vestiges which mark the survival of organs which man no longer needs but which were essential to some lower forms of life which humanity has recapitulated, so does each new stage in religious evolution perpetuate those less-developed stages from which it has emerged. It could not be otherwise. Religion does not exist by itself, any more than life does. As already has been said, strictly speaking there is no such thing as religion in the abstract. There are only people functioning religiously, holding religious ideas and customs, and incorporating them in religious institutions. I cannot help feeling that the recognition of this very simple fact would relieve some who seem to be greatly concerned to rehabilitate the medieval realism. We have long since passed from thinking of scientific law as doing anything or as being anything except a generalization drawn from experiment and observation. We no longer speak about the state as an entity existing apart from legislators and governors, and the other machinery of what we call the body politic. Similarly it is time to realize that when we speak about religion we are speaking about the activities of real people acting and reacting in very real social situations from which institutions, customs, and programs evolve.

Now, real people are vastly interesting subjects of study, and no less so because of their inconsistencies. Sometimes we complacently speak as if in the political field the modern man was quite delivered from the crudities of primitive societies; and yet we lynch criminals and plead the "higher law" for acquitting murderers. It is difficult not to see in such actions the recrudescence of the state of mind of primitive social groups. So, too, in our economic life we cling most vigorously to the formulas of competition when, as a matter of fact, with a rapidity that we deliberately refuse to recognize, we are legalizing a conception of collective bargaining that gives the lie to *laissez-faire*. It is not abnormal therefore to find that religious people, even very intelligent religious people, include in their religious thinking and practices some of the elements which were once the dominating characteristics of a religion in its more simple stages.

1. Reference has already been made to magic, but we find non-magical survivals of primitive religion in all stages of religious development. In fact, a superstition may fairly be described as a vestige of some element of religious experience which has come over from a stage in which it was essential to a religion. One might almost say that to be superstitious is to suffer religious appendicitis. There is no cure for it but surgery. Thus there are women who would not dare say their children are unusually well without touching wood, and there are men who would hesitate to be one of a party of thirteen at a table. Who would think a wedding complete without rice-throwing? What baseball club does not have its mascot? But all of these simple-minded practices which presumably intelligently religious people practice are the survivals of some ancient religious custom of our far-away ancestors.

2. Then, too, religious institutions perpetuate, though generally without the knowledge of their devotees, elements of earlier types of institutionalized religion. The pious Mohammedan still ties rags to trees to remind genii and saints of his prayers and their duties. Even where a religion develops freely many early elements survive. Modern liberal Judaism presents striking illustrations of such phenomena and Christianity has possibly even more striking ones. Indeed, so far has the recognition of religious survivals progressed that, if certain tendencies in a modern theological world were to triumph, religion would have to be regarded as little more than history. But barring these extreme views, even a superficial knowledge of history enables one to know that the Saturnalia festival is preserved at Christmas time. In fact, much of the cult in

any religion is composed of customs, the original meaning of which has been forgotten and which have become sacred or symbolical simply through age. Recall by way of familiar illustration the robes of some of our clergy which perpetuate the dress of the ancient world. The ritual of the Roman Catholic church is particularly rich in such survivals. If these vestiges are not regarded as sacrosanct, they are not without their æsthetic value. Some men will chant creeds they would not otherwise repeat. Almost any religious ceremony is enhanced by this means of linking the modern world with the great course of human history. In fact he would be a most impracticable iconoclast who would ask the complete elimination of cult from a religion or any institution that stands for the conception of the continuity of human experience.

3. In our religious thinking, these survivals and particularly those intellectual forms which have been derived from social experience, play an important and not always a harmless rôle. As has already been stated, the controlling theological ideas of practically all religions were shaped in the great creative period in which local gods become national and a national religion passed on to monotheism. The monarchical analogies are those which show most pertinacity. In fact, in our modern world there are few men who have thus far deliberately undertaken to set forth a theology that shall embody the changed conception of man's relation to the universe itself. It is, however, altogether unfair to think that such experiments have not been made, and are not being made. Religious thought is not nearly so anachronistic as those who know nothing about it appear to suppose. It is true that it can never be quite as precise as a doctor's thesis, but it is making an honest, and I venture to say intelligent, effort to justify and systematize religion from the vantage ground of our modern experience and world-view. Religious thought can

hardly be expected to reshape itself at the behest of every man who has his particular theory to champion. It, like scientific views themselves, will shape itself slowly, in common with the movement of the social mind; but such reshaping, as truly as such a movement, is already in progress.

Yet, in this retranslation of the situation in which religion is involved because of its appropriation of elements of new social experience, we find our thought, and to some extent our experience controlled by the survivals of the monarchical type of religion. In this we are at one with men generally, but with this exception: no other type of religion has ever been held by a civilization as industrial and complicated as ours. It is almost impossible, therefore, at this point, to classify other religions with Christianity. Its nearest species is an academic Brahmanism and neo-Buddhism. In the former the ideal is perfectly distinct, namely, to eliminate and to raise the soul in contemplation as far as possible into the region of the impersonal or at least non-individual. Neo-Platonism somewhat in like fashion attempted to bring man to the Heavenly Vision by ecstasy, but was never able to free itself from the control of survivals and was handicapped by an empirical psychology that frustrated its search for its own ideals. Neo-Buddhism, as it is emerging in the universities of Japan, is a restatement of moral ideals common to all highly-developed civilizations, under the impetus of Christianity. But for an insistence upon vestiges in vocabulary and thought that come from Japanese Buddhism it would be very difficult to distinguish one of these modern Buddhists from the radical Christians and liberal Jews who form the ethical culture groups of America.

4. It will be apparent, further, to any student of society that religion without institutions is of small significance. Religion apart from an institution has not succeeded, any more than a state has succeeded without political institu-

tions. If religion is to be socially effective, its institutions
—and for Christianity this means those of the church—
must be adapted to the changing social order. There are
men who are by temperament anarchic optimists. They
believe that institutions are a hindrance to society, and it
matters little whether those institutions are those of state
or those of religion. Yet even such transcendentalists form
societies of anarchists in order to make anarchy effective.
By the same token the man who wishes to make religion a
purely individualist matter is not without justification for
the maintenance of such a personal luxury, but he over-
looks the fact that in a world like ours religion always has
and always must find social expression, and on both its
intellectual and its institutional sides must partake of social
evolution. So it has come about that religious institutions
are in process of evolution as truly as are religious conceptions. Mohammedanism itself begins to feel the effect of
our modern world and, now that it has broken with political
autocracy, is likely within a generation or two to break with
that religious autocracy which we call fatalism. The Roman
church has not only those new Humanists, the Modernists,
but it is already seeing that in its struggle with socialism it
must adopt the methods of the settlement. The movements
in Asia and particularly in Japan among the non-Christian
religions, though not as marked, are none the less of the
same general type. The ancient Chinese education has been
abandoned and modern textbooks are being introduced
throughout the empire. While it is true that it would be
a little difficult to regard Confucianism as more than a
system of ethics, it can hardly be doubted that the adoption
of the Western school will have decided results in the
case of those religious survivals like ancestor-worship
which Confucianism embodied and preserved. Protestant
churches are already passing through rapid changes as the
social aspects of religion and the social, not to mention the

medical opportunities of the church as an institution are becoming more apparent.

Thus everywhere social evolution finds expression in an evolution of religious thought and institutions that perpetuate vestiges of simpler and earlier stages. Inevitably such a process is accompanied with struggle, for religious survivals are always a conservative force. Just what will be the outcome of this struggle between the representatives of different stages of social experience in religions only the future can tell. But of one thing we may be sure: there will be no cessation either of the impulse to come into helpful reconciliation with a personally interpreted environment, or of the utilization of social experience to justify, control, enrich, and systematize such impulse. And just here lies the pressing task of the religious community. For our modern world needs to be reconvinced that religion is more than a survival, and that the appeal to the universe in terms of personalism is justifiable after concepts inherited from less complex social experience have been abandoned.

THE STRUGGLE BETWEEN RELIGIONS FOR THE SURVIVAL OF THE FITTEST

In what has been said it must have become evident that a distinction is to be made between religion as a functional psychological expression of life, and a religion as a group of beliefs and rites by which this attitude of mind is conditioned and given social expression. The former is as generic as life; the latter is as specific as organisms. The history of religions makes it evident that no one of them can persist unchanged as regulative in a civilization whose moral ideals are superior to its own or whose scientific achievement makes the inherited religious interpretation of existence as a whole outgrown. When a religion has thus found itself out of sympathy with the growing social environment two results follow: either it has been supplanted by another, as was the case in the Roman Empire when the gods of classic mythology were replaced at first in part by Mithraism and later entirely by Christianity; or the religion has adjusted itself in some fashion as has already in a general way been described, reducing its outgrown elements to vestiges, and becomes a new species of religion, as, e. g., in the evolution of rabbinical Judaism from Hebraism. So, too, in the case of the religion of Greece, the simple original Aryan faith was continuously modified by the artistic anthropomorphism of the Homeric literature as well as by Egyptian and Asiatic influences, the worship of the god Hercules, the rise of the Dionysiac enthusiasm and the mysteries, and the work of Æschylus, Sophocles, and Pindar. With the rise of the great schools of philosophy the Greek religion grew extremely complicated, a cross-section of Grecian society showing the existence of the survivals of all the elements which had at some time been locally dominant. Yet the Greek religion did not stop in its philosophical stage but, after the conquest of the Asiatic world, Greece and the Græco-Roman world were invaded by all sorts of Oriental beliefs, and the Greeks by the end of the Christian era ceased to possess any exclusive form of religion. In the place of the sharply-defined national faith of Homeric days, the Græco-Roman Empire possessed a great number of philosophical sects and esoteric religious bodies alongside of popular religion. In it all, however, there was no actual domination of a single religious conception, and classical religion could not withstand the onset of a distinct, unified, aggressive religio-ethical faith like that of Christianity.

If the development of a cosmopolitan civilization thus proves fatal to the more primitive stages of a religion, precisely the opposite is true where a civilization stops at a level set by the religion. The two coalesce. Such, for example, is true in the case of Mohammedanism where the

development of the political and religious concept seems to
have stopped simultaneously at the stage of an imperfectly
moral autocracy. In the break-up of Turkish civilization
which is already beginning because of the introduction of
Western ideals, Mohammedanism will undoubtedly find
itself engaged in a life and death struggle with Christianity
on the one side and materialistic agnosticism on the other.
But such a struggle is not likely to extend far below the
level of those social strata affected by Western civilization.
The great masses of the empire are likely to continue indefinitely under the control of a religion that fits the state
of civilization in which they live. Only as Mohammedans
are educated will Mahomet cease to be the prophet.

The struggle between religions is, then, a struggle not
only between theologies and philosophies but between social
orders. It may occur within a society which, because of economic growth, is differentiating into classes; or it may be
due, as in the case of the Asiatic world, to the introduction
of new social and religious ideals into an older order. From
such a point of view missions became of the utmost sociological significance. Whatever may have been the motive with
which Christians undertook to send missionaries of their
religion to the devotees of another, whether such a motive
were the mere ambition to make proselytes or the genuinely
altruistic motive to save ignorant souls from the punishment of the hereafter, missions to-day are one phase of the
great interplay of social ideals which promises so much for
the future. In the light of the past there can be no question
that changes in the social order will both be conditioned by
and will condition religious evolution, as is strikingly illustrated in Japan. But this change should be sharply defined. The religion best fitted to a social order will not be
a religion foreign to that order. Social history seems to
argue that it is impossible to annihilate one religion by
another. Christianity seriously attempted to replace Ger-

man folk-religion by conventional and wholesale baptisms. But it soon had to make over German gods into saints or devils. Mechanical conversion outside of social transformation is as impossible to-day as in the days of Charlemagne. What really will happen will be a biological development of religions through appropriation and assimilation. As Judaism took up elements from the religions of Canaan and Babylonia; as Greece and Rome appropriated the religious elements of Asia; as Christianity springing from Judaism was transformed by being rethought in terms of Greek philosophy and institutionized in terms of Roman law, so will the nations of the modern world find themselves possessed of religions in which inherited elements are grouped about some nucleating conception into an organic whole. And this organic whole will be the property of those groups of men to whom it is justified by social experience. In other words, a nation will have several religions, although they may be called by the same general name and have many elements in common. Some of these religions will be so unlike those of earlier stages of social evolution as to constitute a new species. In the struggle which comes between these various embodiments of the religious impulse those elements will disappear which are least in harmony with dominating social conceptions of various social groups, and those will survive which are most in accordance with and can contribute most to the development of superior stages of social evolution.

Prophecy is always risky, but unless we utterly misread the present it would seem that there is already emerging throughout the world, under different names, it is true, but none the less essentially identical, a phase of religion the nucleus of which is that of the teaching of Jesus. It is emphasizing brotherhood because of the divine sonship of those who agree with these religious ideals. On the one hand it cannot believe in an anthropopathic God, but on the other it is not ready to deny personality in terms of purpose and reason to the great process in which mankind finds itself involved. Its sympathies are social rather than individual, and its theology is based not on metaphysics of the Godhead interpreted by human analogy but on those judgments of value and those undeniable facts of science which seem to condition all self-expression.

Thus the vanishing point of religious history is still evolution in the sense that the conscious organization of social institutions and thought to bring humanity into helpful relationship with that environment with which it finds itself involved and which men treat as they treat persons, will continue to be a phase of social life. Religion, like any vital reaction, is as real as environment and humanity. But the particular phases of religion and the modes of controlling the expression of this generic impulse are parts of social history. That religion which best enables the religious impulse to express itself in its increasingly complex social environment will survive all others. Other religions will not altogether disappear but they will become vestiges in the more highly developed religious life. And, in this struggle of religions to express religion, Christianity in its ethical and theological sense is certain to be a dominant element.

CHAPTER XVI

SCIENTIFIC METHOD AND RELIGION

By SHAILER MATHEWS [1]

THE preceding chapters have sketched our modern knowledge of the world of nature and man, as well as the methods by which human life has been protected and enriched by better adjustment to its natural environment. It is an extension of method when we also think of religion as an inherited social fact the evolution of which we have traced as an attempt at adjustment to environment in personal ways for the purpose of satisfying personal needs.

The distinction between personal and impersonal relations with nature is comparatively modern. Primitive society knew all but nothing of it. Agriculture was a form of religion as truly as prayer. Indeed, primitive men, unacquainted with chemical and biological forces, did not separate religious practices from tilling the soil and gathering crops. But as civilization proceeded men ceased to believe that good crops could be assured by religious dances, festivals, fasts and sacrifices. Impersonal methods replaced religious. Investigation replaced prayer; fertilizers and soil-testing, magic; irrigation and dry farming, sacrifices. And the same was true of other human interests. Religion was increasingly relegated to the field of personality. In our own day the distinction between the two fields is all but complete.

---

[1] Professor of Historical Theology, University of Chicago.

Yet the growth of knowledge has not destroyed religion. Never has religious activity been greater, and never has it been so intelligent. Despite the attacks of its opponents, despite the ever-increasing complexity of the modern world, it still is an object of primary attention. At the same time science also shows power to enlist the vicarious services of men who, rejecting the temptation to grow rich, devote themselves to human welfare.

Such a statement is of necessity general, but it at least justifies an attempt to discuss in some detail just what the harmony between these two spheres of interest may be, and what service science, as distinguished from the materialistic or mechanistic philosophies of some scientists, may render faith.

SCIENTIFIC METHOD FURNISHES EMPIRICAL TESTS FOR RELIGION IN ITS HIGHER FORMS

Religion, in its highest forms, posits a superhuman personal element in man's environment. In justification of such belief philosophy has endeavored first of all to arrive at some all-inclusive substance of which attributes could be predicated; and theology has started with God as the existence revealed in nature and the Bible. The two methods have ruled most religious systems that have emerged from reflective thought. Having established the existence of a God outside of human experience, it has been the habit of theologians to use biblical conceptions in setting forth the content of Christian teaching.

The difficulty with this procedure has always been seen. Starting with an omnipotent, omnipresent, omniscient Being, the burning questions of religious philosophy have been moral. How could such a Being be good and permit evil? Revelation, it is true, has been drawn upon at this point to show him a Father, but the questions of a theodicy have been the vanishing point of religious thinking. They have been answered, if at all, by the authority of the church and Bible.

Over against this *a priori* authoritative method of expounding the meaning of religion is that of Science. As Professor Ritter has shown, this is empirical. Its first step is observation and experiment; its second, the organization of a working hypothesis; its third, the testing and rectification of such an hypothesis by further observation and experiment.

At this point, however, one caution needs to be emphasized. However readily generalization follows from scientific knowledge, scientific method is altogether different from the philosophical views of scientists. Science is no more naturalism than religion is dogma. To know how rats act in a labyrinth confers no right to dogmatize concerning the mental powers of the man who constructed the labyrinth. A knowledge of biology does not in itself make a man a reliable authority in the field of psychology. The ability to use the spectroscope does not assure infallibility in pronouncements as to the existence of God. With all his attainments, the scientist is human, subject to human prejudices and temperament. The philosophy of a scientist has value only in so far as the scientist is a sound philosopher. Knowledge given us by scientific method is the assurance that experiments under identical conditions will give the same results. Such a conviction is expressed in formulas or "laws." Philosophy is an attempt to correlate these laws or to interpret them in accordance with some general principle.

Scientific knowledge is thus a combination of experience, —often that of others—experiment and faith. It is a synonym for assurance sufficient to warrant action. If there is no confidence that identical experiments will produce identical results, there can be no formulation of law, and no science—simply a host of unrelated facts. Some sort of unity in knowledge sufficient to warrant further experiment is the inevitable outcome of an empirical method. Experiment, hypothesis, testing of the hypothesis by other experiments, reaffirmation or restatement of the hypothesis as a valid basis for further experiment—these constitute scientific method. To such a method religion may well submit its claims.

Method Replaces Authority

1. The substitution of scientific method for reliance upon authority is characteristic of our modern religious thought. The *fides implicita* of ecclesiasticism still persists, but it is no longer a court of appeal for those who have come under the influence of prevailing educational methods. For this rapidly growing body of men and women the methods of science are more conclusive than is authority, for authority itself is in question. A religion, as a social institution in which personal attitudes are expressed, has too frequently identified search for divine help with its own rites, beliefs, and other customs of the social group holding the religion. From such a point of view, it is almost inevitable that one should judge religions as either true or false by the appeal to authority and antiquity. The habit of looking dispassionately at objects of study has helped overcome this religious chauvinism. When religions became the object of scientific investigation, as for example when Max Mueller published his "Sacred Books of the East" ; when the religious practices of primitive savagery were compared by painstaking anthropologists; when the psychological and social elements of various religions were listed; religious faith was seen to rest upon firmer bases than inherited authority. One could hardly feel that a truth which appeared in Christianity was any the less true when it appeared in Buddhism; or that the desire for forgiveness was any the less sincere in the penitential songs

of the Babylonians, than in the Psalms of David; or that the mystic meal in which one partook of a dead and risen god was any less sacred to the worshipers of Mithra than to the followers of Christ. Differences between religions, we have come to see, are incapable of destroying identities of religious attitude. Just as a desire for maintaining social order and military leadership found expression in a great variety of organized governments, ranging from savagery to the republic; so a religion, largely composed of social customs has been seen to be a means of accomplishing an end which another religion has sought to reach by different methods. In other words, historical investigation leads not only to a better understanding of one's own religion, but also to a distinction between a fundamental human need and the means by which it may more or less effectively be met.

2. True, such a change in attitude toward religion has been opposed. Religions have been administered by those who found their authority in some source not open to men and women at large. Priestly control could not continue except as the expression of some monopolized relation with the Deity. The representatives of such control in a church or its equivalent have naturally opposed any attempt to account for its basis. Supernaturalism ceases when explanations begin.

Particularly has this been true in the Christian religion. One of the most common arguments for the acceptance of inherited doctrines is that they have been believed, always, everywhere and by everybody. This, of course, is a strong reason why they should be accepted by those who wish to appropriate a social inheritance, but obviously antiquity and universality are no proof that a belief is accurate. As a test of legitimacy they still leave open the question why a belief was originally accepted.

Authority, like metaphysics, has thus found its rôle in

Christianity seriously lessened. When men began to look into the origin of beliefs, representatives of these beliefs naturally took alarm. This alarm was appreciably increased when the principle of evolution was extended into the field of religion. Of evolution as such, neither popes, councils, nor the writers of the Bible were aware. If the world came into existence in the way scientists declare, the opening chapters of Genesis have no scientific standing despite their lasting religious value. The issue thus raised between the scientist and the theologian is that of method in thinking. In the case of any belief, one must begin with a major premise furnished either by authority or by induction. If one chooses the former, all scientific investigation is at an end. If one chooses the latter in religion, a new method has been introduced and the organized inheritance of dogma, rituals and organization is subject to investigation and inevitable revaluation. It is no wonder that theologians should have been alarmed at the entrance of scientific method into the field of religion. Nor was this anxiety altogether unjustified. The attitude of scientists like Huxley could hardly be called pacific. The unwillingness of the mid-Victorian churchmen to accept natural selection was met with vigorous denial of the possibilities of miracles, the existence of God, and immortality. The conflict between science and religion was not merely the dream of rhetoricians.

This is not to say that all scientists became atheists or militant agnostics. There were many of them who, like Asa Gray, maintained their religious interest and insisted that their scientific experiments did not interfere with their religious faith. Men like Le Conte made real contributions to the harmonizing of religious faith and evolution. Speaking generally, however, religion and science were held by apologists to be in two different planes of life. Such a view means hardly more than that science and religion ought to be completely independent of each other, that the method of one is not the method of the other. Science according to this view could deal with knowledge and religion with values.

But such a position has never satisfied inquisitive religious thinking. If men are to believe in God while their minds testify only to impersonal mechanistic forces, it requires no power of prophecy to see that a truce between the two elements can be only short-lived. If faith does not ignore knowledge, in so far as methods of discovering knowledge prove successful in the one field of investigation, will they tend to be adopted in the other?

3. To say that the empirical method of understanding religion is effective is not to say that religion has surrendered to science in the sense that religious men have adopted the philosophy which scientific men like Haeckel have favored. Men do not grow omniscient even by scientific method. It is to make evident that religion can be regarded as a series of experiments extending across thousands of years and involving a vast number of accumulated actions and convictions. *It is not a philosophy, but a mode of vital action.* Instead of starting with a metaphysical postulate we trace in religion humanity's empirical search after larger and more personal life.

Nor is a scientific method identical with the subject matter of a science. To study and test religion scientifically is not to deny it qualities undiscoverable in other fields of reality to which the same methods of investigation are applied. Method is not identical with discoveries.

True, the material world lends itself more readily to the accumulation of controlled experiments than does the world of the spirit, but the difference seems one of degree rather than of kind. Experience "controls" belief. Change the habits of a man or a people and you change their religion. Its new formulas in turn are tested by new experience and knowledge and used for new satisfactions. The value of a religion can be judged by its capacity to satisfy genuine needs of men and women caught up in the struggle for existence and development as free persons in the midst of impersonal forces. The history of a religion is a history of repeated experiments, resulting in the development or rejection of religious beliefs and practices.

4. From this point of view we accept the religious world-view, not because of some authority—although that, too, is an historical datum—but because as the product of human experience it ministers to human welfare and is consistent with other beliefs that express reality. Religion is thus seen to be life in accord with an increasingly rational working hypothesis based on something other than *a priori* reasoning or ecclesiastical authority. It draws its content and its legitimacy from both knowledge and experience. Religious values are a part of our social heritage tested by experiment.

The record of such experimentation and consequent assurance lies open. Religion is the most universal of social facts—an integral part of our social inheritance. There is no question as to whether scientists should invent religion. It is already in existence. Men have always been religious as they have always been formers of communities and states. The only question which science can fairly raise is as to whether religions have been and are delusions with no standing in the courts of reality.

It is of course to be admitted that many religious beliefs and practices seem unworthy when read in retrospect, but judged in the light of their times they are understandable. The history of religion is the record of the development of values which when followed became permanent. Origins are not ends. The social significance of a religion is to be seen in primitive tribes as truly as in more complicated civilizations. In it a community has deposited, as it were, its supreme ideals and from it has drawn its final sanctions. When religions have ceased so to function, civilizations have lost morale and have succumbed to untoward circumstance. Religions have always been among the conservative forces of social control. When they have completely satisfied longings and needs they have repeatedly checked social progress, but when their forms and beliefs have responded to changing conditions, when they have aided the search for larger life, they have guided human peoples as well as conserved the heritage of real values. In the case of a highly developed religion like Christianity it is impossible for the student of history to deny its power to preserve and develop social values, ideals and practices. Let us not overlook the cruelties of churchmen. Persecution and religious wars are in the pedigree of the Christian church. But there is something more. Think only of the crisis which arose when a great and brilliant civilization disintegrated, cities disappeared, literature was destroyed, works of art were buried, political institutions were abandoned, citizens were massacred, and hordes of armed immigrants inherited a land they had conquered! Yet that was the crisis in civilization which the church had to face in the fourth and fifth centuries. The Dark Ages and the brutality of the Middle Ages are not chargeable to the church. They were the result of social forces which the church had to withstand and transform. And despite all difficulties, it did its work. The only learning was in its circles, the only social ideals were in its teaching, the only social service was in its institutions. But hardly had it made possible the wonderful thirteenth century, when the discovery of America and other causes brought about a complete dislocation of economic, political and educational life. Again Christianity had to face a crisis such as no other religion has been able to face. Again the church survived and gave direction to the Renaissance and revolutionary period of the sixteenth century. So, too, when in the eighteenth century the stress of the new industrial life in Europe and America brought the middle class to power. This period of revolution was not merely political; it was still another shifting of the entire perspective of life. At the start it seemed as if Christianity was to give way to some sort of illumination, or philosophy, or proletarian impracticability; but great religious movements like those of the Methodists, Baptists, and Evangelicals of the Church of England produced men, cultivated attitudes of mind, and organized social agencies which lie at the bottom of much of the social welfare program of modern times.

In all this activity the church has not relied merely on social technique. It has had power to minister to social needs because it has stood for the cosmic reason working in evolution, for a realm of ends, for God in human life, for dynamic morals rather than social conventions.

And what is true of religion in general is true of religion in the individual's experience. For the test to which we subject a religion is our own experimentation as to its ability to meet and answer the problems set by our own life's needs. A philosophy may meet our intellectual disquiet and yet leave us restless and afraid. As will be later emphasized, religion is active, an adjustment and ordering of life, not a formula. It clothes itself in formulas, to be by them protected or impeded, but it is its own justification. Whether ignorant or learned, whether naïve or scientific, men have partaken, as they believed, of some heavenly food and have been satisfied.

Its Value Not the Only Justification of Religion

1. This argument based upon the recognition of the values developed by religion needs reënforcement. Crouching at the very door of such affirmation is the suspicion that the satisfactions religion gives are but inheritances from a

simple age, unsuited to our modern world. Granting that the religious practices and beliefs of mankind have had social value, does our increased knowledge of the universe tend to deny the permanency of such value? If the answer be affirmative, the extinction of religion among intelligent people is inevitable. For if religion rests ultimately upon ignorance, it is only a matter of time when ignorance and religion will be dispelled. Thus new knowledge challenges inherited values. The explanation of religion as a means of conserving social values threatens to be unconvincing. To believe that there is no reality beyond that which is socially created; to admit that the attitude of faith and the practice of prayer are only forms of inner self-communion; and to hold that there is no personal reality in the universe in any way answering to our faith, is to make religion dependent upon one's personal tastes. It is improbable that after one has come thoroughly to believe that the universe is essentially mechanistic, one will seriously attempt even conventional religious activity; it is also true that the religion of personified social values is at one with naturalism in that it denies the possibility of affirming the existence of objective personal elements in the universe. One philosophy sees mechanistic impersonality and the other sees humanity. The fact that one speaks of naturalism and the other of God does not destroy their agreement in denying the position of historic theism.

2. It is protested that the conception of God as a sort of transcendental Uncle Sam is not atheism, and in view of such denial it would be unfair so to characterize it. But this concession simply means that the terms "theism" and "atheism," like "reality," have been redefined. Religion becomes a form of social passion and nature is without personal content except that furnished by humanity itself. The fact that one may urge another to be identified with this community of social interest to which the name

God has been given may serve temporarily, as did its predecessor, Positivism, to stimulate refined minds, but it will
not serve others. When one becomes disillusioned as to the
existence of some being which corresponds to the concept
God, one will lose interest in the concept and in the practices to which it leads. An empty revolver will function
effectively only as long as the highwayman thinks it loaded.
The reason as well as the emotions needs satisfactions. And
back of such personifications as Uncle Sam lie lands and
peoples and other "existential" realities.

3. A further and more basal question therefore remains.
Granting that religion has been able in the past to meet
human needs, are the needs themselves outgrown? Magic,
fetishes, ancestor-worship, the deliverance from a physical
hell, have all passed into the limbo of unneeded satisfactions. The needs they satisfied have disappeared. How
much more secure, men are asking, are the other religious
values? Are they like Santa Claus to be relegated to children and the simple-minded? Utility creates a reliable pre-
supposition of the legitimacy of religion as a social inheritance, but it is not enough. It cannot in itself satisfy the
need of life with what philosophers might call an onto-
logical God. It serves, of course, to corroborate the working
hypothesis of such a God and to encourage men to hold
fast to such a belief. As between the naturalistic interpretation of the universe and the pragmatic conception of God,
the latter has apologetic worth, for it is one thing to live in
a world without personality and another to live in a world
of personal values. There is undoubted help in appealing
from Kant of "Pure Reason" to Kant of "Practical Reason." God in himself can never be known, since all knowledge must be of the phenomenal, but His existence must be
assumed as necessary for our moral life. Moral values certainly imply moral realities in the universe. Yet, unsup-
ported by other considerations, this argument seems to me

to leave the world without a complete basis for the religious faith it avowedly needs. Of course, honesty would demand that even if there is no ground for believing in some personal existence within the universe, we should live bravely in a universe without God. Bitter as the conclusion would be, it is more honorable to live honestly than to live self-deceived. If we have to grant that science has given the victory to naturalism, let us recognize our conquerors. At all costs, let us be honest.

We seek, therefore, to go further than pragmatism. Granting that the direct, *a priori* arguments for the existence of God land the thinker in antinomies, and that therefore we must admit a reverent agnosticism as to infinity, can the faith demanded by our moral life find support in our knowledge of the universe? Imperfect as that knowledge may be, does its growth justify rather than weaken religion? While it may not directly prove the existence of God, does it disclose in the universe that which favors the religious world-view and warrants faith in a God of things as they are and are becoming? If religion is to abide, our new knowledge must reënforce the argument from the contribution of religion to the permanency of personal and social values. Religion as a social inheritance subject to the laws of social psychology and institutional development must find its place among the legitimate interests of the modern man. If science cannot demonstrate God by its experiments, can it justify the religious expression of life?

CHAPTER XVII

SCIENCE JUSTIFIES THE RELIGIOUS LIFE

By Shailer Mathews [1]

STARTING this with religion as a social inheritance involving definite attitude, experience and world-view, *our method requires us to discover whether it is tenable in view of the growing knowledge of reality given us by science.*

Let us again notice that the question is not directly whether science can demonstrate the existence of God or whether it will permit the institution of an already existing social institution. It is simply whether a social inheritance, rich in power to do well or ill, has rational grounds for further existence. Have we still a right to be religious in the full sense of the word? Is a personal attitude toward the universe permissible in the light of our knowledge?

The answer to such questions is not doubtful. Science warrants religion because it affords evidence of immanent reason, purpose and personality in the cosmic environment and its discovery of the laws of human life.

1. *The new conception of matter has ended the old materialism.* Instead of matter composed of minute dead, solid pellets, subject to external forces, the physicist gives us something so subtle as to be almost beyond our power of imagination. To common sense matter seems without motion; to the physicist it is activity itself. At the risk of too confident statement, it would seem that the ether, the existence of which is as yet the commonly accepted work-

---

[1] Professor of Historical Theology, University of Chicago.

ing hypothesis, has within itself vortices or undulations which constitute atoms. Within these atoms are outer negative and inner positive charges or electrons. Matter is thus formed by the aggregation of an infinite number of vortices within a hypothetical ether, each vortex carrying negative or positive electrical charges. "Solidity" is thus a phase of activity.

By the side of such a description of matter the religious concept of cosmic spirit does not seem irrational. The existence of God can be inferred from experience as truly as the physicist posits ether. And for the same reason. The phenomena of experience can be unified and explained only on such grounds.

Now it is one thing to have activity as the vanishing point of our knowledge of the material universe, and another to have "dead" matter. The chasm between the spiritual forces presupposed by religion and the activity presupposed by the physicist is certainly far less difficult to bridge than that between the older contrasts of spirit and matter. If our interest were in metaphysics, we might well pause at this point to ask whether such a view of the universe tends toward pantheism. There certainly is a kinship between the two conceptions of cosmic reality. But religion is not dependent upon metaphysics. Like any aspect of life its concern is its own activity. Just as the processes of assimiliation proceed in the living organism, whatever may be the ultimate nature of matter, so the process of religion proceeds whether the ultimate philosophy be pantheistic, pluralistic, or theistic.

It is certainly no small gain for religion to discover that the farther one penetrates into the mystery of the physical universe the closer does one come to self-directing activity. Certainly spirit is no more impalpable than an electric charge when it seems as if there is nothing to charge. If it be urged that the material universe is more susceptible to tests by experiment than is the spiritual, one can reply that the distinction is in subject matter rather than in method. The religious world-view is no less a rational working hypothesis than is the hypothesis of universal activity in the ether. Both alike are to be tested by their efficiency to explain phenomena, serve to unify knowledge and enrich life.

2. *Astronomy is helping religion in that it is forcing us to believe that this ultimate activity, is, so far as the word means anything, infinite.* There is nothing knowable beyond it. True, the infinity given by astronomy is bounded by agnosticism as to the unknowable. But despite its caution science here gives us far more imposing and realistic concepts than metaphysics. To reach the negative conception of that which is "not bounded" means far less than to have reality piled upon reality until the mind is incapable of thinking further. Our knowledge of astronomy is giving us new concepts of boundless reality. Whatever the philosopher may decide is the nature of space and time, whether or not they be concrete, the impression given by a study of the abysmal universe is the nearest approach to actual infinity of which the human mind is yet capable. And that infinity is activity, itself the stuff from which things that act are made.

It is almost startling to find that the physicist and the chemist find no new sort of matter in this universe of the astronomer. The activity of which atoms are constituted is everywhere. The discovery of new elements in the sun or other heavenly bodies even when not followed by the discovery of the same element in nearer areas of space discloses no sort of structure different from that already found in the hydrogen atom. Nowhere is there anything that could be called dead matter. Everywhere from the utmost bounds of a super-galaxy to the ultimate nucleus of an atom is ceaseless activity.

Of this infinite activity all things consist. Of that there can be no doubt. As to its ultimate nature we are yet ignorant, but as to its way of operation we are daily growing wiser. Two characteristics are constantly more in evidence; this infinite activity in whatever form it is found, either in the atom or in the constellation, first, *is describable in general statements called laws* and, second, *shows within itself appreciable tendencies*. True, this description is as yet incomplete. But science has not yet repudiated its faith that such limits are set only by our ignorance. Order is the first law of science as truly as of Heaven. Natural law has sometime been described as analogous to the statutes of a criminal code. Law is said to ''do'' this or that. From such a point of view law would be not altogether unlike the Fate of the classic mythology in accordance with which all things, even the gods themselves were forced to act. But scientifically speaking this use of words is misleading. A scientific law is a statement of the results of observation, a description of the way forces act rather than a cause of such action; an interpretation of activity, a formula of similarities.

To the lay mind as well as the scientific there is something overpowering in this conception of infinite activity capable of being understood by human minds. But if possible even more overpowering is the fact that only as it is regarded as possessing of intelligible unity is it in any way understood. When one speaks of the laws of thermodynamics or gravitation one is not talking about something which compels matter to act in a certain way; he is describing the operation of the atoms themselves. Until infinite activity is seen to contain within itself something self-directive akin to reason the possibility of scientific investigation is denied. I am not now attempting any discussion of the ultimate relations of the rational with the physical activity. Words here become hardly better than symbols. Pantheism, monism, dualism, what are they but more or less successful attempts to set forth the fact thrust upon us by every scientific experiment, that throughout infinite activity there is that which is akin to human reason? Without the assumption that this activity is, as it were, sane, it would be idle to make experiment. Mathematics and hypothesis would lead only to more confusion.

Science, therefore, unconsciously, yet inevitably, assumes and so discloses rationality within the cosmic activity. Whatever may be the metaphysical explanation of the phenomena, we can never take any other attitude towards the physical world. Wherever there is physical activity there is intelligibility. But what is this but to say that the primary position of religion is warranted? To be religious is to give full weight to this rational element. It is first of all to live in accordance with the universe as capable of being understood, in the process of unfolding itself to human understanding. There must be intelligence in an intelligible universe.

3. *Within the field of observation and experiment science is showing us not only laws and order, but processes and tendencies.* Ultimate origins and destinies are beyond our knowledge, but within the reach of human understanding we find not only activity, but direction. One cannot read even superficially the history of science without seeing the continuous disclosure of this better understanding of a universe composed of that which is ever becoming. But tendency and process are more than the "becoming" seen by the ancient Greeks. Nothing is more genuinely the gift of our recent sciences than the discovery of the orderly processes in the universe. It is, of course, what might have been *a priori* assumed from our knowledge of infinite activity describable in laws, but how much more significant does it appear when we trace in the nebulæ the history of our universe and in the rocks the history of our earth!

Our inability to understand fully the origin of species does not prevent our perception of the process itself. As Professor Newman has already stated, there is no other hypothesis which so well unifies and makes intelligible the facts at our disposal as that of evolution.

1. It has already been pointed out why men should have said that there was a contradiction between the acceptance of the hypothesis of evolution and religious faith identified with a theory of the inspiration of the Bible. Such objections, however, seem negligible in the light of the actual contribution to our religious thinking made by the hypothesis of evolution. If there is tendency, development, process, evolution, then the infinite activity is working toward ends. True, we make such affirmation with caution. We do not dare to make too anthropomorphic any explanation of what we observe in nature. But one thing seems beyond peradventure: the mechanistic conception breaks of its own weight when one studies any process. If the fittest to survive are to survive, whence comes the concept of fitness? If there be no reason or purpose in the expression of life, why should not the unfit survive rather than the fit? Such a question of course sounds imbecile, but the mere fact that it is unthinkable shows how legitimate it is to see in the realm of activity something akin to purpose. Language, we admit, is here a poor medium of expression. Our vocabularies inevitably gravitate towards either abstraction or personification. The "entelechy" of Driesch or the "dominants" of Reinke "which sway whatever energies are available just as men use tools" are simply ways of saying what the human mind instinctively affirms, that nature reveals a kingdom of ends as well as of histories. Scientists themselves are more or less consciously drifting into similar interpretations. Even a polemic for materialism sees that "the more we learn about the nexus of natural phenomena the greater becomes our power to prophesy future events,"

and can speak, though rather inadvertently, of a "goal."
There is more effective argument for the existence of reason
and purpose in the universe in the evolution of humanity
than in the older argument from design. Indeed, it may be
said to be the developed modern form of that argument.
For science discloses a universe of activity characterized by
traits so analogous to what we call reason and purpose in
human beings, as to be unintelligible unless such qualities
are recognized.

2. The working hypothesis of evolution thus carries us
to a most important conclusion. For it must take into
account two facts: human personality and the dependence
of human evolution upon a contributing environment.

I do not understand the scientist to argue that the total
universe is evolving in the same direction. We are not yet
in a position to affirm universal progress, although that,
too, does not seem beyond conjecture. In certain areas, indeed, there appears to be disintegration or at least mere
transformation of physical forces, as when motion is transformed into heat and light. But whatever may be the
philosophic and scientific limitations of a view like Berg-
son's, elemental logic makes it clear that there is nothing
from which human personality could be derived unless it
be that activity which constitutes ultimate existence. It is
true that we cannot yet define with definiteness just what
the term personality indicates in humanity. We must
grant its genetic relationship with animal instinct and behavior. No one, however, is in serious doubt as to what it
symbolizes and represents. Even the champions of the
mechanistic interpretation of life and the most extreme of
behavioristic psychologists know the difference between
themselves as champions of a philosophic theory and the
animals and plants on which they base that theory. We
do not hesitate to say that the process of chemistry is impersonal, and that the mental, volitional, and intellectual

life of the chemist is personal. It is this distinction that makes the mechanistic interpretation of life unsatisfactory. Men deliberately undertake to be different from what they are. They judge the worth of some present state in the light of something more or better which they may apprehend. Discontent leads to conscious readjustments. Conditions are not only understood, they are evaluated in terms of that which is helpful or hurtful. It may be that this power of value-judgment which belongs so far as anybody knows only to living animals of the species *Homo sapiens* is, as Professor Herrick says, a function of living matter. But "living," as he points out also, is a term of vast and unexplored potentiality. To speak of living matter is something very different from speaking of dead matter, or merely matter. Neither the biologist nor the sociologist would deny the existence of thought, nor that the more complicated the living organism the more complete is its intellectual and volitional action. To speak of a human being as a "self-stoking, self-repairing, self-preserving, self-adjusting, self-increasing, self-reproducing machine" is to jumble together contradictions. As Professor Thomson says, "It has inside of it a human thought." A machine is not capable of knowing that it is any particular sort of machine. If it were not possible to trace the development of the human species into personal self-expression, the full significance of the evolutionary process could never be known. Suppose, for example, human evolution had stopped with *Pithecanthropus erectus*. It would still be true that the increase in the complication of physical structure paralleled increase in self-directive and purposeful elements of life, but all of the phenomena and activities which the term personality connotes would never have existed. It is only at the appearance of *Homo sapiens* that life reached a degree of organization possessed of self-conscious and self-determining power capable of making a study of itself and of seeking non-physical values. Personality is not omitted by any psychology that moves beyond the observation of rats and birds, behavior and nerve-reflexes.

The other consideration is of equal importance. Evolution is not merely a mutation of elements existing in lower forms of life. Species *b* is not merely another species *a* from which it was evolved. The conservation of energy has to do only with the impersonal side of activity. Each new species of life discloses an addition of characteristics. While it would, of course, be incorrect to think of these additions as simply quantitative, it would be equally impossible to think of the new species as identical with its predecessor. Evolution does not proceed in a vacuum. Whatever other forces may have been at work in the production of species, certainly the environment is not to be ignored. Whether it be conceived of as making quantitative additions which are assimilated by the organism or as conditioning evoked and controlled changes implicit in the germ plasm, the environment is in the evolutionary process. Every new advance in species means some additional feeding in of influence on the part of the environment. The new situation in which the new species is found involves the stimulus or the contribution of the environment as much as the organism.

This fact must be correlated with the appearance of personality in life. If life has grown more personal while organisms have grown more complicated this must be due in part to the influence of an environment within which there must be that which can evoke personality in the progressive series of organisms. This is an immediate corollary of the fact that the evolutionary process is conditioned by the environment with which the organism is in dynamic relation. Air has to be presupposed for the transformation of a water-breathing animal into an air-breather. There was no need or possibility of such evolution before our atmosphere existed. Similarly, as human evolution leads toward human personality and this process must be within and dependent upon an environing universe, there must be something correlative to it, something capable of assimilation by it within the environment. Otherwise personality would never have appeared. *It is impossible to think that personality could evolve from the exclusively impersonal.* There must be that in the environment which the personality-possessing organism can appropriate, and with which it can act harmoniously. If something analogous to personality were not in the universe the processes of evolution would have stopped below personality as we know it in man.

From such a point of view the evolutionary process of humanity can be described as a successive development of organisms sufficient to appropriate or respond to personal elements from the environment, the influence of which is argued by the process itself. And this is what would be expected from our discovery of elements in the universe analogous to reason or purpose.

It is at this point, of course, that one is again tempted into the all but forbidden field of metaphysical speculation. For if the ultimate of reality is activity, and if we see it in certain lines of process up through what we call matter into life and then into personality, it is easy to see why naturalism fails to satisfy human needs. It fails to take into account all the qualities which observed evolution shows must have been implicit in ultimate activity, namely, those which could produce personality, and so must themselves be not foreign to personality. In other words, we might think of this ultimate activity as being, so to speak, potentially dualistic, mysteriously capable of expressing itself in ever-enlarging personal as well as impersonal ways, environments and organisms.

But be the value of metaphysical speculation of this sort what it may, when one sees activity characterized by reason and purpose and at the same time producing strains of living matter (no longer a word of materialism) which in turn has personality as a function, one has come in sight of God. For by these facts the great conviction is forced upon us that just as we men and women have personality within that mass of chemical and physical activity which we call our bodies, so in the infinite universe of activity is there immanent an infinite Person whose existence and character account for its relationships, its tendencies and its achievements. Such a Person cannot be static perfection, but creative of new values far in advance of chemical compounds. The analogies with which we most satisfactorily think of Him cannot be derived from that which is mechanistic or static. Whatever may be the shortage of our vocabulary, and however misleading may be a literalizing of analogies and symbols, no such terms can so well express this ideal and value-producing activity of the universe as those which we use in description of humanity itself. We can no more think of ultimate activity through the analogy of the atom alone than we can think of it through the analogy of the bull of the Egyptians. Personality must be described by the analogies of persons.

4. Here we see further legitimacy given religion by science, not as a mere philosophy of value or of ideals, but *as an active, personally adjusted living with the value-producing elements of the universe*. Science has certainly made it plain that an organism which has resulted from its relations with a certain environment must live in accordance with that environment if it is to maintain its distinctive characteristics. A water-breathing animal must live in water, an air-breathing animal must live in the air. Otherwise it ceases either to exist or to maintain its characteristics. Religion is the application of the same law to human

personality. As the animal must live like an animal rather than as a crystal in a universe which has produced him, so a person must live personally in a universe from which he springs. Only thus can his evolved characteristics be maintained. Religion from this point of view is certainly more than a mere vestigial survival of primitive customs; it is a law of life. He who lives in accordance with the laws of his own being will live in sympathetic, help-receiving personal relationship with the environment upon which he depends. To live in any other wise is to run counter to the laws of life itself. We are thus not merely using conventional terms when we say that a man can lose himself. He loses his personal self whenever he tries to live as if he were impersonal, just as a plant would lose its life in conditions in which a crystal could continue indefinitely. Thanks to our new knowledge we not only are assured of the legitimacy of the religious world-view but are urged to the actual process of religious living. Though we may not know directly the metaphysical object of our faith, life in accordance with faith is reasonable and imperative.

In brief, therefore, we can see that science is contributing not only a method of testing the legitimacy and value of religion, but also it is giving us thought-patterns with which one may set forth its nature. (For we always live in new conditions by the use of patterns drawn from experience already possessed.) When the highest reach of experience was that of the state, men organized their lives with the universe imperfectly but to their best possible advantage through the pattern ideas of the state. But now as science has enabled us to get glimpses of reality, of cosmic process and of life, it is giving us far more inclusive and improved pattern ideas. These pattern ideas of science enable us effectively to coördinate with other realities the urge to protect and enrich life by seeking help personally from the environment in which personalities find themselves.

CHAPTER XVIII

SCIENCE GIVES CONTENT TO RELIGIOUS THOUGHT

By SHAILER MATHEWS [1]

IN the light of conclusions thus far reached, we can see how religion as a social inheritance can be made to function more effectively by means of an enlarged knowledge of man's relation to the universe. We are not in a universe either inactive or without personal forces. God is as much implied in a scientifically tenable religion as in the more primitive. Modernism in religion is no synonym of naturalism. It may not so sharply speak of the "natural man" and the "regenerate man" as does dogmatic theology, but its reserve in using such terms is not due to any failure to distinguish between non-religious and religious attitudes. Such a distinction, which is the heart of the older thought, is of the very essence of religion. Man's progress is due not to isolated but to divinely environed and influenced life. God's transforming presence is indispensable for personal development. But such influence is itself subject to new understanding.

I. For, most important of its services, *science is giving new content to the conception of God.*

1. How the idea of God has been shaped by human experience as known through anthropology has already been shown. It is necessary here only to call attention to the chief characteristics of the process by which the modern idea of God has been produced. Speaking generally, the

---

[1] Professor of Historical Theology, University of Chicago.

conception of divinity has been reached by transcendental-izing something regarded as a source of power upon which man feels dependent. It has been expressed in terms which were drawn from social life and social technique. The primitive savage conceived of setting up personal relation-ship with his environment through magic, or some unreflect-ing coercion, flattery or gift. As civilization developed, men came to think of tribes as having their independent gods, much as they had their independent ancestors. So the god was pictured with the magnified power and attributes of a supreme and unconquerable tribal chief. In the case of the Hebrew race, after several centuries of social devel-opment, the conception of a tribal god was expanded by the finely endowed minds of prophets into that of the only God. The questions as to His ultimate power which were set by national defeat served in the prophetic teaching to argue purpose and ends which could be reached only through the moral life of humanity in general and of the Hebrew people in particular. Thus there developed wholly outside of the field of philosophy the spiritual and moral monotheism which is the unique contribution of the Hebrew religion to human life. Parallel with, but to all intents and purposes independent of, this development, we see the com-plicated civilizations of China, of India, and of the Hellenis-tic world reaching noble conceptions of a Supreme Being, by the road of reflection and philosophy. Popular mythol-ogy, while not destroyed, no longer satisfied daring thinkers like Plato and Aristotle and Zeno who saw a Mind in nature and its activities. Religious faith in Greek culture, how-ever, found its finest expression less in religion than in philosophy. In the Christian community these currents of Hellenistic philosophical monotheism and Hebrew empirical monotheism combined in the Christian doctrine of the Trin-ity, that quintessence of the religious history of thousands of years. Its duplicate is not found in any religion. The

triads of the eastern religions are not the trinity of the
Christian faith, for the Christian faith is an attempt to
answer philosophically the problem set by experience as
to how monotheism is consistent with varieties of religious
experience. Other religions have developed alongside of
their philosophical speculations an apparatus religion which,
as in Hinduism, may include millions of gods. The Christian movement was beset by the same temptation, especially
in view of its faith in Jesus as the divine Savior. But the
keen minds of the early Christian theologians refused any
approach to polytheism, even though it were the subtle
polytheism of Arianism. It combined the Christian experience and the Greek philosophy in the pattern already
furnished by the law courts. God was conceived of as a
single substance possessing three *personæ* or self-revela-
tions and relations. Therefore, whether met in nature as
the Father, or in Jesus as the Son, or in the human experience as the Spirit, He was always God—not three, but one.

2. But such a metaphysical conception has always been
too refined for the needs of religion. Men might make it
the test of their orthodoxy, but when speaking religiously
theologians have always thought of God in the terms of
politics. He was a sovereign who gave statutes to his subjects, punishing their disobedience, rewarding their loyalty
and forgiving them their sins. As political institutions
rose and fell and rose again in Western Europe the conceptions of God were continually enlarged. For Christian
theology has always been more than a philosophy. It has
been the practical religious formula of those creative social
minds and groups which produced western civilization.
Indeed, in a way almost startling, it is possible to trace the
reappearance of political experience and ideals in theologies. Roman imperialism, feudalism, nationalism, the imperfect constitutional monarchy of England, capitalistic
control of modern civilization, all have furnished patterns

by which God has been set forth in his relations with humanity. These pattern ideas still hold sway in the minds of the majority of Christians, but they are increasingly seen to be imperfect analogies. The doctrines derived as corollaries from them are increasingly foreign to the new social mind being shaped by science. It is inevitable, therefore, that, particularly among the young, there should be need of a new pattern idea by which the religious urge can be coördinated with the other thing which is known to be real. Instinctively men feel that they have not outgrown God while outgrowing kings.

This new organizing concept is not to be found in metaphysics but in that reciprocal relationship of the universe and the atom, the environment and the organism to which science is accustoming thought. "God" is infinitely more than "sovereign," and religion is more than transcendental politics. Just as truly for religion is God more than the Absolute. Personal relations demand personal agents. Prayer is always the practice of those who live the religious life, and prayer is the utilization in religion of conversation, the one great means of adjustment between persons.

But how shall one pray to a Personal Being when the concept of reality given by science is activity and process in a universe to which science can give no bounds? The contradiction will probably always be unsolved. The apparatus of thought is incompetent to construct reality beyond realities. The proposals of pluralists like James to solve the difficulty by making the God of religion finite behind whom there is some vast abyss of Being seems to me even less satisfactory than those given by a proper understanding of the Christian doctrine of the Trinity. Why should the modern man with his growing sense of the oneness of reality revert to a euphemized polytheism? Let him live bravely with the God he has. The fundamental difficulty lies in the habit of thinking of God through highly refined but static materialistic analogies. Only as we draw analogies from our own vital activity and think of Him as ever seeking new values through personal self-expression do we get any suggestion of relief from the intellectual impasse into which speculative thought throws us. For we can then think of process in values without inferring incompleteness. Yet even thus the impasse in thought is not wholly removed. There must always be something beyond our knowledge in religion, however truly the adventure of faith is directed by knowledge.

3. But as has been said so often, religion does not wait upon philosophical solutions. Religion is a form of life itself, and from our experience with persons rather than with matter we must draw our creative analogies. "Life" and "person" as well as "king" and "emperor" furnish us patterns for our thought of a relationship with God. It is one of the inestimable merits of Jesus that he himself made vital rather than political terms dominant in his teaching. To him God is a Father rather than a sovereign, and fellowship with Him is life rather than political obedience. To think of God as infinite personality working out—on this earth at least—values nobler than those already attained, is to think in sympathy with Christ and science alike. The God of the philosopher is a problem. The God in whose infinite activity we live and move and have our being, whose self-expression environs us in the eternal process toward personality which science is helping us read, is, to speak reverently, a greater God than mankind has ever worshiped. The faith which such a conception evokes requires no renunciation of knowledge, no belittling of man. It deepens reverence, heightens awe and awakens courage.

4. It may be questioned, however, whether science justifies belief in the God of love. And it must be frankly admitted that whoever thinks of God as immanent (if such

a word is permissible) in infinite activity, hesitates to speak easily of His will in the details of life. The simpler faith that can find explicit purpose in this or that event—punishment in earthquakes, rewards in prosperity, chastening in sorrow—seems too adventurous. Like Job, he who recognizes the cosmic power of God is apt to bow his head in awe of that which passes understanding. But he none the less can believe in a Good Will. And this faith is more than that seen when men distrusted the actual goodness of their deity and so attempted to win His favor by gifts. The religious man of intelligence cherishes no such misapprehension. He knows that the God of a universe like ours is not to be cajoled, wheedled or bribed. He must be either good-will or ill-will. The materialist, of course, immediately protests that it is impossible to speak of moral qualities in the universe, and, therefore, even to raise the question of such qualities is illegitimate. To such objection there is no answer if the general position of naturalism be tenable and the universe is no more than chemical and physical forces. So long, however, as one sees reason, purpose and personality not only in humanity but immanent in the universe, and the capacity to organize ideals is felt to be an element of human evolution, so long will there arise the question as to whether there be love in the universe as we know it.

All students of philosophy know the difficulty confronting such a question. The problem of evil, arising from the existence of that which causes suffering, is one before which thinkers have ever stood in despair. So long as the conception of the deity was drawn from human passions such philosophical questions were not so pressing. It was enough to say that the natural misfortunes, like labor and death, were punishments justly inflicted on disobedient subjects; but as men thought more philosophically such an explanation grew unsatisfactory. Theodicy, that is, the effort to justify the goodness of God, became one of the most bitterly debated questions. It would be a mistake to say that no progress was made in this discussion. Men came to see more clearly that they were not without rational grounds for accepting the biblical conception of God as good and loving. But these considerations have been greatly reënforced by our new knowledge of process within the universe. For whatever limits may be set by knowledge, there is evidence of a process from the more simple to the more complex and in one line, at least, toward personality. So far, therefore, any fair interpretation of evolution parallels the conception of God's love. We may not and usually cannot find specific evidence of purpose and love in specific experiences. But a process towards the more personal cannot be regarded otherwise than as arguing good-will. Misery caused by impersonal actions and the suffering which comes from perverted personality must be judged from this point of view. If there be that which is evil, there is process to that which is better. Many an evil is an outgrown good; many another is an incentive to knowledge and social service. If we do not have grounds for optimism we are certainly justified in holding to what Professor James calls "meliorism." And religion, exploiting this fact, looks forward to an ever more intelligent faith in the goodness of a God of that which is becoming. The love of a God of cosmic law is immeasurably grand. To live like Him as good will is the supreme opportunity of our confused humanity. And however great our ignorance, it is not the part of either reason or courage to doubt that the knowledge yet to be given by science will give us new glimpses of Eternal Love.

5. I do not discuss the question, so favored by many thinkers, as to whether God himself is still "becoming." I am content to say that He shares with us our own becoming just as any environment must share with the organism

it is helping evolve. He continues the creative process through humanity's too often tragic efforts to build up the new personal environment of society. If logic seems to force the conclusion that God is growing greater, I do not trust the major premise which such logic assumes. It seems only a new and subtler form of anthropomorphism. Agnosticism as to that which lies beyond knowledge and co-operative adjustment to that which is rationally probable, seems safer and more reverent. Of this, however, we can be assured: in the universe we know and within which we live there is Personal Activity sufficient to give meaning to infinite activity and strength for our ethical tasks. That is the God religion needs. If the logic of metaphysics is not satisfied, religion as life is unabashed. For action and development are more than logic, which too often is the last and spectral refuge of a faith that dares not trust itself.

II. *Biology and the newly developed medical sciences give more realistic meaning to the conception of sin as the violation of divine will.*

1. As long as the deity was regarded as the super-king, the importance of formally keeping His statutes and particularly of maintaining proper customs of worship was paramount. The general tendency in all religions has been for views of sin to pass from the region of morality into that of cult. Ecclesiastical organizations have naturally emphasized the importance of those elements of a religion like doctrines and rites over which they have peculiar control. At the same time they have too often been complacent regarding the performance of anti-social acts by those who were in good ecclesiastical standing. Some one has rather bitterly said that if the church had treated sinners as it has treated heretics, we should have a better world in which to live.

With our enlarged understanding of human life no such

conventionalizing of sin is possible. In fact, the term itself is being replaced by words of more concrete import. To violate the will of God immanent in the process from which we have sprung and are continuing is something more than working on the Sabbath day or eating meat on fast days or even keeping the Ten Commandments. It is a violation of the very situation in which life must be lived. We know what is the law of God for humanity because we can watch its operation from the fire-mist to our own day. It is that men shall grow less animal, less mechanistic, more personal, more regardful of others' personality, more possessed of good will. For the individual to violate this great law unintentionally is to bring suffering upon somebody; for him to violate it intentionally is to bring not only this suffering, but also guilt, that is to say, an injury to his own personality which must bring future loss and suffering.

And this conformity to the actual will of God as seen in the operation of the universe extends over into the entire realm of human interest. We work according to the will of God when we use natural forces in accordance with their true nature and to further personality. Heredity expresses this will no less than gravitation. By faulty matings a race may become degenerate. But conversely, we are coming to see how the total character of mankind is not altogether dependent upon its own decision; so intimately connected are our personal and impersonal elements that much which once seemed to be evil is really misfortune. The germ plasm reacts to stimuli; the individual to causes of behavior. What we once thought was mental inattention on the part of children we now know may be due to adenoids or defective vision. Criminals are often undeveloped personalities given freedom which is beyond their power to appreciate. Nor is that all. Human personality is motivated, not merely by the future with its values, but by the past with its instincts. Our bodies abound in vestiges of organs that once had functions and now are merely survivals, like the vermiform appendix and the tonsils, capable of injury but not of help. We inherit instincts, reflexes and mental habits from our animal forebears. Many of them are characteristics which we hold in common with the animals and the more primitive man. In themselves they may be neither good nor bad; when they are given their proper position in our lives, as in play and recreation as well as normal physical acts, they are good. When there is added to them an element of responsibility set by the demands of our own developing personality for going on to higher personal values, moral character attaches to their use. When these inheritances of the development through which humanity has passed become dominant and check the forward-looking personal elements, degeneration follows. When a man yields to this backward pull of outgrown goods he is running counter to the immanent law of the universe to progress. Suffering and loss are bound to ensue from such maladjustment, and one aspect of the tragedy is that such maladjustment may bring suffering to others than its author.

2. It may be that this fact, now given us so clearly by our knowledge of evolution, is what Augustine had in mind when he spoke about original sin. The evidences which he gives of the corruption of human nature are precisely those passional elements which humanity shares with other animals. So far from being corruption, they are inheritances carried along according to the observed operations of life as a basis and apparatus of personality. Practically all great moralists have seen the same conflict within the human soul that Plato and Paul describe, but the absence of any scientific knowledge of the origin of the human body and the development of human personality led them to very imperfect and sometimes grotesque explanations of the facts. But it is not hard to see how sympathetic Augustine

might have been with our modern knowledge of evolution and eugenics.

3. Similarly, our better understanding of the processes of social evolution enable us to appreciate how sin belongs to the realm of social life. We have long since ceased to speak very confidently about the social organism, but with our knowledge of archæology and history we can see how a group carries along within its own institutions elements which it has outgrown, and yet which are liable to reassert themselves. Society, like human beings, needs to transcend its origins. Indeed most of our social evils are outgrown social goods. Like slavery, they once marked an advance in community life but each in turn has become hostile to further advance in the development of human personality. The old antithesis of the theologians between the Augustinian and the Pelagian doctrines of sin is thus resolved. Sin, whether it be found in the individual or in the social order, is yielding supremacy to an outgrown good. Such an attitude is rebellion against the great law of the development of the human personality and a misuse of our social environment.

III. *Science helps give fuller meaning to the conception of spiritual regeneration.* 1. If sin be maladjustment to immanent purpose that brings suffering, salvation is adjustment with consequent progress and happiness. By virtue of his derivation and dependence upon his cosmic environment man can assimilate God. Such a conception continues the values of the Christian doctrines of reconciliation, but expresses them more literally. Its accord with the concept of life is far more trustworthy as a pattern for religion than is the conception of sovereignty with all the appurtenances of a monarch's rage and difficult justification of repentant rebels. To be saved is to become, one might say, convalescent. The deteriorating effects of maladjustment to an environment are offset by a new and more intense personal adjustment to the conditions set by the presence of personality in the universe in which we are. And such a readjustment, of course, is that set by more than worship. Wherever men come in contact with personality must they make personality supreme. We can no longer be indifferent to perverted animal inheritances making diseased bodies through abnormal relation with the natural world and so hindering personal self-expression; or to social conditions and social orders that preserve in humanity's very environment forces that make suffering easy and good-will difficult. All these are to be overcome. The moral life is a phase of the total process. We must live in harmony with our neighbor as truly as with God; society must be prevented from being the source of evil influence. For as the impersonal order has built up an impersonal environment, so personal forces in humanity have built up a personal environment. With all of these must one live in coöperation, and through the social environment must come new uplift and help. We can separate between morality and religion only in the sense that morality regards human personality and religion, divine personality. The attitude of mind and basis of relations are the same in both. In morals, as in religion, there is development. Right and wrong are not abstractions, but sign posts of experience pointing toward good-will as the basis of proper human relations. The origin of moral distinctions is a matter of speculation, but that morality, society and personality must be contemporaneous is now beyond question. It seems to be probable that at least one factor in the development of a morality is the empirical discovery that one sort of action brings trouble to a group and another sort brings peace and happiness. Therein is revealed a structural law in human development. No society can survive where stealing, murder and adultery are universal. The respect for property rights became a foundation of primitive social order and an element of morality in the nature of the case. The history of humanity has been from the simple towards the complex. *Mores,* customs are means of transforming individuality into personality. And these must be in dynamic adjustment to cosmic will.

2. Society like the individual has obviously its impersonal environment and beliefs. Geography is no small factor in the development of *mores*. But there has also been in humanity a constant spiritual direction in development. Men in society have passed from a reliance upon force to reliance upon constructive *mores*. There has been increasing tendency to treat men as possessed of inherent value other and beyond economic capacity. Gradually men are seeing that duties are more final than rights; that good-will is more productive of social and personal well-being than acquisitiveness and coercion. All this argues that human society is giving increasing weight to elements of personality. As men learn to live together personal freedom increases with mutual dependence. This seems paradoxical, but the evidence is all about us. How much more mobile is modern man than is the savage, how much less at the mercy of natural forces, how much enriched by the division of labor, how much readier to recognize the rights of the weak, of women and children!

Such considerations are what might be expected from our general knowledge of the cosmos. If the tendency in human evolution is away from the impersonal to the personal, the course of social evolution such as we actually observe is just what might have been expected. A strictly mechanistic universe could not be imagined to contain such developments within the range of humanity as have actually taken place within the last twelve or fifteen thousand years. Morality might, indeed, be described as the precipitate in custom and social sanctions of the gains made by the human response to the personal element of its cosmic environment.

It is a long way from the pack to the nation, from instinct to social organization, from slaughter to self-sacrifice. But such progress is the projection of that process which science enables us to chart. As the material world formed new environments for its own process, as the living matter made new environment for life, so personality in the social process makes new and personal environment for personality. To live in harmony and coöperating endeavor with this new environment so that personality may be more complete and self-determinative is to be righteous. For such living is in accord with that Will which evolution has revealed and embodied in the newly created environment of society. Through it as through the immeasurably older environments of life and matter an immanent God is working.

3. It would be idle to say that life in accordance with the higher elements of cosmic and social environment is without its struggles and its difficulties. Pleasure always tempts men to make an outgrown good supreme. Water runs more easily downhill than uphill. It requires less struggle for human life to live according to a lower good already enjoyed than to face the future and struggle for the attainment of goods which are to be gained through faith in their worth. But struggle for new values is the imperative of personal development. A universe in which there is an everlasting process towards higher values is no place for cowards. Men of violence must ever take the kingdom of God by conflict. But whoever approaches this great fact with the help of scientific conceptions can believe that in this struggle within the moral order, as truly as in the process which has produced the moral order, personality is not without assistance from environing reality. As a plant that reaches for the sun is helped by the sun, so human beings and human societies who seek to live according to the higher values of personality find themselves assisted

by the forces that have produced personality. God is not static and men cannot grow by remaining passive. Religion is simply another way of setting forth the promise and potency of evolving life by help-gaining relations with God, formed directly by individuals or indirectly through society.

4. Here we find new suggestions for answering the problem set by humanity's hope for immortality. Religion, especially of the higher type, has emphasized the world beyond death, but too often its hopes have been so naïve as to be untenable by those who have come to know more accurately the structure of the human personality. And it must be admitted that the realm of affirmation regarding the future life grows less picturesque and concrete as knowledge grows. The astronomer finds no place for a physical heaven in the universe, much less a physical hell. Our human souls are something other than breath which comes from our nostrils to take shape in some material fashion in the skies. The difficulties arising from the physiological aspects of modern psychology are undeniable. Its very vocabulary seems foreign to those who have learned to speak of "self" and "soul." But so long as *living* matter differs from all other, so long as consciousness is irreducible to chemical symbols, no weight can be given to objections against the persistence of human personality after death drawn from the dissecting room. We cannot expect to find evidences of life in a corpse. Nor are we justified in omitting facts in the interest of precise formulas. At the risk of using unwarranted speculation, we cannot stop our estimate of what life may be until we have gone beyond the molecules of matter to that mysterious ultimate activity in which eternal reason and purpose are truly present. In the production of individuals may there not be conservation of cosmic personal activity as truly as of impersonal energy?

At all events, one cannot with assurance stop short of the possible projection of the development of personality and of individuality across the cessation of consciousness in what we call the body. Who knows the full content of life? The production of the individual rather than of species (which at best is only a term of similarities in individuals) seems to be, as Professor A. P. Mathews has so well said, the road of evolution. And if the human individual develops his personality through his capacity to believe in and follow immaterial, timeless values, it would seem as if in evolution scientific knowledge led to a major premise for faith in immortality. The force of analogy drives us to the belief that death is something more than the maintenance of the conditions set before death. It seems rather the occasion of still further development of those pioneer qualities of the eternal activity focused in our individuality. That is certainly possible enough to be a basis of hope. Whether scientific investigation may ever develop apparatus and methods which will press beyond the veil of our own bodies into the life of non-animalistic personalities one cannot say. It is hard to see why experimentation in that field is not as legitimate as in any other. One thing, however, seems certain, a life which has sloughed off animalistic survivals through death cannot be of the same order or possessed of the same interests as a life where universal activity makes mind a function of living matter as we know it in the brain. The kingdom of heaven, says Paul, is not eating and drinking, but love and joy and peace in the Holy Spirit. To that great conviction I cannot help thinking our conception of the origin and growth and operation of the human spirit offers no final objection. At the least it is a satisfactory working hypothesis so long as ultimate physical reality is not "dead" matter but activity.

5. The Christian theologian may move forward with

this same central teaching of science in his representation of the values of the Christian religion. "Life" gives him more tenable premises than "sovereignty." He can find new understanding of Jesus as a revealer of an infinite and immanent but self-expressing personality. He gets a new sense of the participation of God in the sorrows and struggles of humanity with outgrown goods expressed so picturesquely but imperfectly in the various doctrines of the atonement. He will understand the Bible as an inestimably valuable product and record of man's response to his true environment expressed in his growing discovery of God and of God's gradual self-manifestation to man through personality and society. In a word, he will come to see that those fundamental values of the Christian religion which lie behind the pattern ideas in which they have been expressed by theology are not opposed by an interpretation of the universe which science is helping us gain. But that interpretation must take into account all of the facts at its disposal, the personal as well as the impersonal. It can be only those who are ignorant either of the real nature of Christianity or of science or of both who will find antagonism between the two spheres of expressing the meaning of ultimate Activity. Our scientific knowledge justifies our religious faith when it must assume an intelligible, order-producing process. The greatest contribution which our religion makes to man is its adventurous and indomitable assertion not only that the Godhead can be known from the things which are visible, but that the values which are invisible to the microscope and the telescope are eternal; that the universe of life is a universe disclosing good-will; and that evolution when properly understood as a coöperative process of the impersonal and personal elements within the universe leads to religion as its highest phase.

CONCLUSION

THE PERSONAL EXPERIMENT OF FAITH

We reach thus the end of our study, not in an absolute knowledge of that which lies beyond experience, nor yet in some truth that is beyond doubt, but in the right to live religiously. Religion is more than knowledge. It is life itself, adjusting itself to the mystery of its environment, gaining from that experience new confidence and development and erecting the new and more personal environment of society. It can therefore never be identical with science, nor is it to be subsumed under science. Action is more than an accumulation of facts. Nor does religion wait upon science. It antedates science. Needs of body and of personality are imperative and cannot wait for a knowledge as to how they can be fully satisfied. Religion may never hope to explain the cosmos, but men can live religiously in the cosmos with self-respect as long as our growing knowledge of its operations shows religion to be rational and science itself partakes of a kindred faith.

It is just here that the opponents of religious faith overlook man himself. No science is competent to explain all facts. Precision of generalization must not be reached by omitting recalcitrant facts. A humanity that is only a group of mechanistic and chemical reactions is not the humanity we know. Love, faith, aspiration, discontent, courage and religion are activities which must be included in our understanding of human life. Life, like philosophy, might seem easier if such disturbing forces were absent from our daily tasks. Men have retreated from human interests to the cultivation of a meditative life which would ultimately plunge them into unconscious passivity or lead to a heaven beyond death. But when one comes to estimate forces which have raised humanity from the all but animal life of the old Stone Age into the richly furnished world of interests in which we now live, it is easy to see that struggle towards these spiritual values has been the directive force of such recent progress as humanity has actually made. Had our ancient ancestors been content to remain passive, they would, like their neighbors, the beasts, never have started their descendants upon the road to our modern life. Humanity, if its physical weakness had permitted it to exist, would be still living in caves, suffering from cold and heat, without art, education, literature, or political institutions. Only a perverted mind could prefer such a life to that of a modern man, or judge history to be without progress.

This struggle against the impersonal forces of nature towards larger social control in the interests of freedom and self-direction must be carried on since it is in response to the immanent Will. However men assailed by lassitude and discouragement may be tempted to yield to what fate may give, it is the part of courage and of self-respect to choose a life that seeks more personal ways. And such a life is its own best justification. It finds and welcomes evidence that it is not inconsistent with the universe in which its great adventure lies. It would adjust itself ever more intelligently to reality. But it will hold itself second to none. It is more than an adventure. *It is a creative experiment carried on in coöperation with the God of that which is ever becoming.* Unlike science, it does not undertake simply to know, but to transcend knowedge by the active creation of values that are beyond physical discovery. The facts germane to such discovery are precious, but religion will not be bound by the past experiences or by scientific agnosticism. Its goal is not an explanation of the universe, *but a life with that reason and purpose and personality which science increasingly warrants it in assuming in the universe.* And this adventurous experiment finds its justification not only in the heightened worth of lives who make it and the more personal environment they create in society, but in the new knowledge which science gives. Whatever may be the limitations of our understanding, however much we may be oppressed by the vastness of the universe, however mysterious may be our own existence, and however poignant may be our sense of failure to understand ultimate reality, we are a part of all that we see, fellow-workers in the process which is carrying life away from its biological and chemical inheritances toward the half-glimpsed, but ever-approached freedom of the completed personality.

INDEX

Neo-Buddhism, 371
Neolithic man, 213
Nitrogen, in plant life, 309
Numbers, 256 *sq.*

Paleolithic man, 213
Personality, cosmic implied by evolution, 399
Phylum, 183
Physiological psychology, 247 *sq.*
Piltdown man, 201
Pithecanthropus, 186, 199
Planetesimal hypothesis, 102
Planets, 64
Plants, scientific study of, 141 *sq.*; primitive, 144; first land plants, 149 *sq.*; vascular, 151 *sq.*; seeds, 153; sociology of, 155; method of growth, 306; breeding, 314; diseases of, 315
Pottery, manufacture of, 221
Prayer, 406
Preliterate peoples, 215; culture, 229
Primitive man, 212 *sq.*; theories regarding, 223
Progress, doctrine of, 218
Protoplasm, properties of, 127
Protoplast, 146
Psychology, physiological, 247; experimental, 249; abnormal, 258; limitations of, 262

Regeneration, spiritual, 413
Relativity, theory of, 59
Religion, a, defined, 3

Religion, nature of, 2 *sq.*, 352, 378, 401; evolution of, 351–377; scientific method in, 379, 381; a series of experiments in living, 384 *sq.*, 389; implies more than values, 387 *sq.*; tenable in the light of science, Chap. XVIII; not dependent on philosophy, 407
Religions, primitive, 358

Salvarsan, 277
Sanitation, development of, 323
Satellites, 66
Science and religion, conflict of, 1 *sq.*; 6; 106, 383; not contradictory, 11; medical, science and religious faith, 283; caution concerning, 380
Science, defined, 19; assurance in, 380
Science, method of, Chap. II, 23, 171; tasks of, 243
Serum, 343
Sewage, disposal of, 334
Sex impulses, 259
Sex in plants, 147 *sq.*
Siblings, 174
Sin, idea of, 410
Social evolution, older theories of, 215; tested, 218; presuppositions of, 219; supposed ages, 222; diffusion of cultures, 223; facts of, 227 *sq.*; periods of, 229 *sq.*; true conception of, 233 *sq.*; formula of, 235
Solutrean man, 213

(1)

THE END